Albrecht Zimmermann, James Ellis Humphrey

Botanical Microtechnique

Albrecht Zimmermann, James Ellis Humphrey

Botanical Microtechnique

ISBN/EAN: 9783744662208

Printed in Europe, USA, Canada, Australia, Japan

Cover: Foto ©berggeist007 / pixelio.de

More available books at **www.hansebooks.com**

BOTANICAL MICROTECHNIQUE

A HANDBOOK

OF

METHODS FOR THE PREPARATION, STAINING, AND MICROSCOPICAL INVESTIGATION OF VEGETABLE STRUCTURES

BY

DR. A. ZIMMERMANN
PRIVAT-DOCENT IN THE UNIVERSITY AT TÜBINGEN

TRANSLATED FROM THE GERMAN BY
JAMES ELLIS HUMPHREY, S.D.

NEW YORK
HENRY HOLT AND COMPANY
1893

Copyright, 1893,
BY
HENRY HOLT & CO.

ROBERT DRUMMOND, ELECTROTYPER AND PRINTER, NEW YORK

PREFACE.

THE methods brought together in the present volume are, of course, chiefly taken from the literature scattered through various original papers and text-books. But the author has always endeavored, so far as possible, to reach an opinion from his own experience concerning the methods described; and many of the details and modifications of previous methods contained in the book are due to his own investigations.

However, the literature used has been as fully quoted as possible at all points, in so far as it seems of present value. Works of merely historical interest are not referred to, since the book is designed only for practical use.

If the writer has overlooked many statements of value, it is to be hoped that it will be understood and pardoned by those familiar with the immense extent of botanical literature, especially in recent years. The author will be grateful to any one who will call his attention to such omissions.

Regarding the quotation of literature, it may be said that numerals are placed after authors' names in the text, which refer to the literature list at p. v, the first (Roman) number indicating the work, and the second, the page of the work cited. Where the author was not able to consult the original in the preparation of this book, the abstracts used are referred to.

In the arrangement of the organic compounds, Beilstein's

Handbuch der organischen Chemie (II. Auflage, Leipzig, 1886-1890) has been substantially followed.

The illustrations, where the contrary is not expressly stated, are prepared from the author's original drawings.

The manuscript was practically completed in July, 1891; but I have tried to include the more recent literature, so far as possible, during the printing.

TUEBINGEN, March, 1892.

TRANSLATOR'S NOTE.

THE need of a good handbook of microscopical methods, as applied to plants, has for some time been evident.

Such as we have had have only partially covered the ground, and are now mostly out of date on account of the rapid advances of the last few years.

The appearance of the original edition of this work very satisfactorily met this growing demand so far as concerned students familiar with the German language; while its evident thoroughness and the familiarity with its subject-matter shown by the author in the selection of the most useful among the innumerable published methods give it especial value for students of less experience.

The belief that elementary students should have access, from the first, to the best methods, and the fact that few such English-speaking students read German readily, have led me, with the support of the present publishers, and with the cordial consent of the author and publishers of the original edition, to undertake its translation.

In preparing this English edition, I have followed quite closely the original. Certain notes and tables have been added which, it is hoped, will add to its practical usefulness to American and English students; and certain matters not included by the author, which seem to demand notice, have been discussed in their proper places.

I am especially indebted to Dr. Zimmermann for preparing, for this edition, notes on several important results of very recent studies; and I have added a few annotations of the same sort, thus bringing the present edition as completely as possible up to date. All additions by the translator are enclosed in square brackets.

WEYMOUTH HEIGHTS, MASS , July, 1893.

CONTENTS.

PART FIRST. GENERAL METHODS.

	PAGE
1. The Observation of Living Plants and Tissues. §§ 1–5	1
2. The Investigation of Dried Plants. §§ 6, 7.	5
3. Maceration. §§ 8, 9	6
4. Swelling. § 10	8
5. Clearing. §§ 11–27.	8
A. Chemical Clearing-methods. § 12	9
B. Physical Clearing-methods. §§ 13–27	11
I. The Ordinary Method of Transfer from Water to Canada Balsam. §§ 14–22	12
II. The Transfer from Water to Canada Balsam without Alcohol. §§ 23–25	17
III. The Use of other strongly refractive Mounting Media §§ 26, 27	18
6. Live Staining. § 28	19
7. Fixing and Staining Methods. §§ 29–40.	20
A. Fixing. §§ 32–34	21
B. Removal of Fixing Fluids. § 35	22
C. Staining. §§ 36–39.	24
D. Fixing and Staining Microscopically Small Objects. § 40	27
8. Microtome Technique. §§ 41–52	29
I. Imbedding in Paraffine. §§ 43–49	31
Ia. Imbedding in Celloidin. § 49a	35
II. The Attachment of Sections. §§ 50–52	37
9. Making Permanent Preparations. §§ 53–62	40

PART SECOND. MICROCHEMISTRY.

A. INORGANIC COMPOUNDS.

1. Oxygen. § 63	44
2. Peroxide of Hydrogen. §§ 64–67	45
3. Sulphur. §§ 68–70.	47
4. Hydrochloric Acid and its Salts. § 71	48

	PAGE
5. Sulphuric Acid and its Salts. § 72.	49
6. Nitric Acid and its Salts. §§ 73-76	50
7. Phosphoric Acid and its Salts. § 77	52
8. Silicic Acid and the Silicates. §§ 78-81	53
9. Potassium. § 82	56
10. Sodium. § 83	56
11. Ammonium. § 84	57
12. Calcium. §§ 85-99.	57
a. Calcium Oxalate. §§ 86-89	57
b. Calcium Carbonate. §§ 90-92	60
c. Calcium Sulphate. §§ 93, 94	62
d. Calcium Tartrate. § 95	63
e. Calcium Malate. § 95a	64
f. Calcium Phosphate. §§ 96, 97	64
g. Recognition of Calcium in the Ash. § 98	66
h. Recognition of Calcium in the Cell-sap. § 99	66
13. Magnesium. §§ 100, 101	67
14. Iron. § 102	68

B. ORGANIC COMPOUNDS.

I. *Fatty Series*.

1. Alcohols. § 103.	69
Dulcite. § 103	69
2. Acids. §§ 104-106	70
a. Oxalic Acid. § 104	70
b. Tartaric Acid. § 105	70
c. Betuloretic Acid. § 106	70
3. Fats and Fatty Oils. §§ 107-112	71
4. Wax. §§ 113-115	74
5. Carbohydrates. §§ 116-125	75
a. Glucose. §§ 118-120	77
b. Cane-sugar. § 121	78
c. Inulin. §§ 122, 123	78
d. Glycogen. § 124	80
e. Dextrine. § 125	80
6. Sulphur Compounds. §§ 126, 127	81
a. Garlic-oil. § 126	81
b. Mustard-oils. § 127	81
7. Amido-compounds. §§ 128-130	82
a. Leucin. § 129	82
b. Asparagin. § 130	82

II. *Aromatic Series*.

1. Phenols. §§ 131-133	84
a. Eugenol. § 131	84
b. Phloroglucin. § 132	84
c. Asaron. § 133	85

CONTENTS. ix

	PAGE
2. Acids. §§ 134–136	85
a. Tyrosin. §§ 134, 135	85
b. Ellagic Acid. § 136	86
3. Aldehydes. § 137	86
Vanillin. § 137	86
4. Quinones. §§ 138–141	87
a. Juglon. § 139	87
b. Emodin. § 140	87
c. Chrysophanic Acid. § 141	88
5. Hydrocarbons. §§ 142–149	88
a. Ethereal Oils. § 144	89
b. Resins and Terpenes. §§ 145–149	90
6. Glucosides. §§ 150–164	92
a. Coniferin. § 151	92
b. Datiscin. § 152	93
c. Frangulin. § 153	93
d. Hesperidin. § 154	93
e. Coffee-tannin. § 155	94
f. Potassium Myronate. § 156	95
g. Phloridzin. § 157	95
h. Ruberythric Acid. § 158	95
i. Rutin. § 159	96
k. Saffron-yellow. § 160	96
l. Salicin. § 160a	96
m. Saponin. § 161	96
n. Solanin. § 162	97
o. Syringin. § 163	98
p. Glucoside (?) from the Stimulus-conducting Tissue of *Mimosa pudica*. § 164	98
7. Bitter Principles. §§ 165, 166	99
a. Calycin. § 165	99
b. Spergulin. § 166	99
8. Coloring Matters. §§ 167–197	100
a. Pigments of the Chromatophores. §§ 168–179	100
α. Chlorophyll-green. § 169	101
β. Carotin. §§ 170–172	101
γ. Xanthin. § 173	103
δ. Coloring Matter of *Aloë* Flowers. § 174	103
ϵ. Coloring Matters of the *Florideæ* (Phycoerythrin). § 175	103
ζ. Coloring Matters of the *Phæophyceæ* (Phycophæin). § 176	104
η. Coloring Matters of the *Cyanophyceæ* (Phycocyanin). § 177	104
θ. Coloring Matters of the *Diatomaceæ* (Diatomin). § 178	105
ι. Coloring Matters of the Peridineæ (Peridinin and Phycopyrrin). § 179	105
b. Fatty Pigments or Lipochromes. §§ 180, 181	106
c. Other Coloring Matters dissolved in Fats or Ethereal Oils. § 182	107
d. Coloring Matters dissolved in the Cell-sap. §§ 183–185	107
α. Anthocyanin. § 184	107

	PAGE
β. Anthochlorin. § 185	108
ε. Coloring Matters which are first contained in the Cell-contents, but later penetrate the Wall. §§ 186–188	108
ζ. Coloring Matters which only occur deposited in the Cell-wall. §§ 189–191	109
g. Coloring Matters which are deposited upon the Cell-wall. § 192–197	112
α. Thelephoric Acid. § 194	112
β. Xanthotrametin. § 195	113
γ. Pigment of *Agaricus armillatus.* § 196	113
δ. Pigment of *Paxillus atrotomentosus.* § 197	113
9. Tannins. §§ 198–208	114
10. Alkaloids. §§ 209–222a	119
a. Aconitine. § 210	120
b. Atropine. § 211	120
c. Berberin. § 212	120
d. Brucine. § 213	122
e. Colchicine. § 214	122
f. Corydaline. § 215	122
g. Cytisine. § 216	123
h. Opium Alkaloids (Morphine, Narcotine, Narceïne). § 217	123
i. Piperine. § 217a	124
k. Sinapine. § 218	125
l. Strychnine. § 219	125
m. Theobromine. § 220	126
n. Coffeine, Theine. § 221	127
o. Veratrine. § 222	127
p. Xanthine. § 222a	128
11. Nitrogenous Bases. § 223	128
Nicotine. § 223	128
12. The Proteids and Related Compounds. §§ 224–239	128
a. Reactions of the Proteids. §§ 224a–234	129
b. Nucleïns. §§ 235, 236	133
c. Plastin. § 237	134
d. Cytoplastin, Chloroplastin, Metaxin, Pyrenin, Amphipyrenin, Chromatin, Linin, and Paralinin. §§ 238, 239	135
13. Ferments. §§ 240, 241	136
a. Emulsin. § 240a	136
b. Myrosin. § 241	137

PART THIRD. METHODS FOR THE INVESTIGATION OF THE CELL-WALL, AND OF THE VARIOUS CELL-CONTENTS.

A. THE CELL-WALL.

In General. §§ 242, 243	138
1. The Cellulose-wall. §§ 244–250	139
2. Lignified Membranes. §§ 251–261	143
3. The Cuticle and Suberized Membranes. §§ 262–272	148

		PAGE
4. Gelatinized Cell-walls, Plant-mucilages, and Gums. §§ 273-282		154
a. Amyloid. § 276		156
b. Wound-gum. § 277		157
c. The Gelatinous Sheaths of the *Conjugatæ*. §§ 278-282		157
5. Fungus Cellulose. §§ 283, 284		160
6. Paragalactan-like Substances. §§ 285-287		161
a. Reserve Cellulose. § 286		162
b. Paragalactan. § 287		162
c. Arabanoxylan. § 287a		163
7. Callus and Callose. §§ 288-291		163
8. Pectic Substances. §§ 292-296		166
9. Ash- and Silica-skeletons of the Cell-wall. § 297		168
10. On the Developmental History of the Cell-wall. § 297 a-e		168
11. The Finer Structure of Cell-walls. § 297 f-p		170

B. THE PROTOPLASM AND CELL-SAP.

In General. § 298		174
1. The Nucleus and its Constituents. §§ 299-348		175
I. The various Methods, in General. §§ 300-330		176
a. Fixing Methods. §§ 300-313		176
b. Staining Methods. §§ 314-327		180
c. Simultaneous Fixing and Staining. §§ 328, 329		188
d. Staining *intra vitam*. § 330		189
II. The Resting Nucleus and its Constituents. §§ 331-339		190
a. Its Recognition. §§ 331-334		190
b. Its Constituents. §§ 335-339		191
III. The Karyokinetic Figures. §§ 340-342		192
IV. The Inclusions of the Nucleus. §§ 343-348		194
2. The Centrospheres. § 348 a-d		198
3. The Chromatophores and their Inclusions. §§ 349-364		201
I. Methods of Investigation. §§ 350-354		201
II. The Finer Structure of the Chromatophores. §§ 355-357		203
III. The Inclusions of the Chromatophores. §§ 358-364		204
a. Proteïn Crystalloids. § 359		205
b. Leucosomes. § 360		205
c. Pyrenoids. §§ 361-363		206
d. Oil-drops. § 364		208
4. The Eye-spot. § 365		209
5. The Elaioplasts and Oil-bodies. §§ 366-370		209
6. The Iridescent Plates of various Marine Algæ. §§ 371-372		212
7. Microsomes and Granula. §§ 373-375		212
8. The Cilia. §§ 376-380		214
9. Proteïn-grains. §§ 381-392		215
a. The Fundamental Mass. §§ 382-384		216
b. The Crystalloids. §§ 385-387		217
c. The Globoids. §§ 388-391		219
d. Crystals. § 392		221
10. Proteïn Crystalloids. §§ 393-395		223

		PAGE
11.	Rhabdoids (Plastoids). § 396	223
12.	The Acanthospheres of the *Characeæ*. §§ 397–399	224
13.	Starch-grains and Related Bodies. §§ 400–415	225
	a. Starch. §§ 400–410	225
	b. Floridean Starch. § 411	230
	c. Phæophycean Starch. § 412	230
	d. Paramylum. § 413	230
	e. Cellulin-grains. § 414	231
	f. Fibrosin-bodies. § 415	231
14.	The Mucus-globules of the *Cyanophyceæ*. §§ 416–417	232
15.	Tannin-vesicles. §§ 418–421	234
16.	The Reaction of the Various Cell-constituents. §§ 422–427	236
17.	Plasmolysis (Plasma-membranes). §§ 428–434	238
18.	Methods of determining whether Certain Bodies lie in the Cytoplasm or in the Cell-sap. §§ 435, 436	240
19.	Aggregation. §§ 437–440	241
20.	Artificial Precipitates. §§ 441–444	242
21.	The Loew-Bokorny Reagent for " Active Albumen." §§ 445–448	244
22.	Protoplasmic Connections. §§ 449–454	245
23.	Contents of Sieve-tubes. §§ 455, 456	248

APPENDIX. METHODS OF INVESTIGATION FOR BACTERIA.

In General. § 457		250
I. The Observation of Living Bacteria. § 458		250
II. Fixing Methods. §§ 459–465		251
1. Cover-glass Preparations. §§ 459–464		251
2. Sections. § 465		253
III. Staining Methods. §§ 466–476		253
1. Loeffler's Methylene blue		253
2. Ziel's Carbol-fuchsin		254
3. Ehrlich's Solutions		254
4. Gram's Method		255
5. Staining Tubercle Bacilli		256
6. Staining Spores		257
7. Staining Cilia		257

TABLES FOR REFERENCE 259
LITERATURE LIST 265
INDEX 285

BOTANICAL MICROTECHNIQUE.

Part First.

GENERAL METHODS.

I. The Observation of Living Plants and Parts of Plants.

1. The direct observation of microscopically small Algæ and Fungi generally offers no difficulties, and is best conducted in the culture fluid of the organism under observation.

In case of delicate objects which would suffer from the pressure of the cover-glass, this may be prevented by the interposition of paper strips, capillary tubes of glass, or similar objects. The method proposed by Kirchner (I, VII) and Vosseler (I, 461) is especially adapted to this purpose. It consists in providing the cover-glass with small "wax feet" at its corners, for which a mixture of wax and turpentine is best. This is prepared by adding to melted wax one half or one third of its bulk of Venetian turpentine, while stirring constantly.

This mass adheres very closely to glass and possesses besides a certain plasticity, so that one can readily use immersion lenses and can easily slide or depress the coverglass.

2. Where investigations are to be continued through a longer period of time, covering the organisms in a small

drop of fluid with a cover-glass in most cases greatly hinders the access of oxygen. In such cases the so-called moist-chamber is commonly used. This is closed above by a cover-glass, from whose lower surface a drop of culture fluid, containing the micro-organisms to be observed, hangs free (observation in the hanging drop).

Such a moist-chamber can be prepared most simply from card-board about 2 mm. thick. A rectangular piece is cut from this, a little smaller than the glass slide to be used, and a square opening is then cut in its middle, with sides about 4 mm. shorter than the cover-glass to be used (cf. Fig. 1, *a*). These card-board cells are thrown, before use, into boiling water, by which they are at the same time saturated and sterilized.

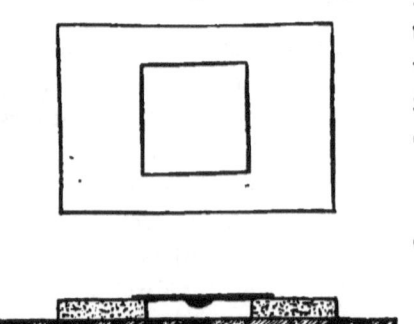

Fig. 1.—Moist-chamber for the culture of micro-organisms.

While still wet they are placed on slides, and then the cover-glasses, with the organisms to be cultivated in hanging drops on their under sides, are placed upon them. The covers are so placed as to rest upon the card-board on all sides, as is shown in Fig. 1, *b*, which represents such a moist-chamber in longitudinal section.

The culture drop is thus freely in contact with a large volume of air, and is protected from evaporation by keeping the card-board wet by the occasional addition of a few drops of water.

Besides this, various other moist-chambers have been used for the same purpose, as, for example, slides with ground cavities, or glass blocks with hemispherical or lens-shaped hollows. These offer, in some cases, certain advantages; but in general they are based on the same principle, and their use is easily understood (cf. Strasburger, I, 415 ff., and Behrens, I, 51 and 162).

Moreover it is usually necessary to renew the culture fluid in all these moist-chambers from time to time, perhaps twice a day. With some larger objects this can be easily accomplished by taking up the drop with filter-paper and replacing it by a fresh one.

3. In many cases one may employ with good results the methods recently recommended by several writers, which permit a continuous change of the culture fluid. I will describe in detail only the method of Klercker (II), which seems to possess certain positive advantages over those of Rhumbler (I) and Schönfeld (I).

J. af Klercker uses in the first place an English slide (cf. Fig. 2) to which two glass strips, L, about .14 mm. thick, are cemented with Canada balsam, as shown in the

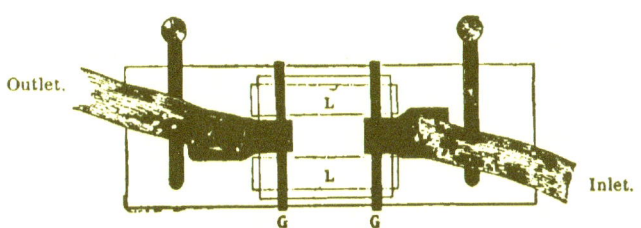

FIG. 2.—Slide for culture in running water. After J. af Klercker.

figure. In the middle of the channel thus formed is placed the organism to be cultivated, and a large cover-glass is laid over the whole. If the capillary space thus formed between the slide, the two glass slips, and the cover is not wholly filled with fluid, a sufficient quantity to fill it is added. Then a strip of linen is pushed under the cover from each side (S, Fig. 2), and the latter is fastened to the slide with rubber bands, G. The slide thus prepared is attached to a second slide with wax, in order to be more easily movable, and then the whole is fastened to the stage of the microscope with two clips, K, in the manner shown in Fig. 3.

The supply of water passes through the siphon, H (Fig. 3), from the larger beaker (B_1), which is protected from dust by a glass plate, by the aid of the linen strip (S_1) drawn through it, whose free end lies upon the linen strip S (Fig.

2.) The escape is provided for in the same way by a linen strip (S_{11}) which communicates with the small beaker (B_{11}). The rate of flow can, obviously, be regulated by changing the height of the water in the beaker B_1, as well as by filling the siphon H more or less tightly with linen. J. af

FIG. 3.—Part of a microscope with apparatus for culture in flowing water. After J. af Klercker.

Klercker commonly allows a flow of about 50 ccm. in 24 hours.

4. In order to protect long-continued cultures of Algæ from Bacteria and other fungi, one may, with Klebs (III, 492), add to the culture fluid .05% of neutral potassium chromate (K_2CrO_4), which does no perceptible injury to Algæ or to sections of higher plants. Palla (I, 322) added for the same purpose .01% of potassium bichromate ($K_2Cr_2O_7$).

5. If one wishes to observe sections of larger plant-tissues in the living condition, they must, naturally, be always more than a single cell-layer in thickness, so that they may contain cells in no way injured by cutting. Most cells, however, die pretty quickly in pure water, when the chromatophores and the nucleus often become completely deformed by swelling strongly.

Therefore very various fluids have been used for the study of living cells, such as solutions of neutral salts, 2 to 10 per cent solutions of sugar, gum arabic, and fresh egg-albumen. In the study of nuclear divisions in the embryo-sac, a 1.4%

solution of potassium nitrate (saltpeter) rendered Treub (I, 9) good service.

The observation of uninjured chromatophores can be very successfully conducted in many cases, as Bredow (I) has stated and the author can confirm, by placing the fresh sections directly in oil (pretty fresh olive-oil). The cells not only remain alive in this for a long time, but the oil acts as a clearing agent, from its strongly refractive property, and usually excludes the air from the intercellular spaces pretty completely.

When using other fluids for observation, however, one must commonly use a filter-pump for removing the air; and sections on the slide can be treated, or larger pieces of tissue can be filled with fluid before cutting. The pieces are placed in small crystallizing dishes and are attached by filter-paper to their bottoms so as to remain wholly covered with fluid during the pumping.

The same object can be attained by placing the sections containing air in boiled water; but in general the pump will accomplish the desired end sooner.

II. The Investigation of Dried Plants.

6. For the investigation of *dried Algæ and Fungi*, Lagerheim (I and II) recommends that they be first softened in water and then placed in concentrated lactic acid, in which they are to be heated until they show small bubbles. The organisms thus treated, having completely resumed their original forms, can then be directly studied in the lactic acid.

7. *Herbarium material of the higher plants* may often be made suitable for sections by simply soaking in water. In many cases the dried parts may be treated with dilute ammonia or caustic-potash solution, which should be washed out before cutting. The strength of the solutions and the time of their action must be governed by the nature of the objects, and must be determined for each separate case. Small or friable objects are best cut with the microtome after being imbedded in paraffine or similar substance (cf.

§ 41 ff.). Objects which have become strongly colored in drying may best be bleached by *eau de Javelle* (cf. § 12, 4).

III. Maceration.

8. In many cases, especially when concerned with the size, form, or structure of the membrane of the various cells, it is desirable to separate an organ into its component cells. This proceeding, which is commonly known as *maceration*, depends upon the fact that the middle lamella which is present between adjoining cells is dissolved by various reagents, so that the cells separate from each other. The ready solubility of the middle lamella depends, according to the researches of Mangin (IV and VI), in most cases upon the fact that it consists of various *pectic* compounds. Thus one can bring about an isolation of the cells in very many objects by treating them first with acid alcohol and then with ammonia (cf. § 295).

Besides these, one may use many other and varied media for the same purpose. Thus it is sufficient in many cases to place the tissues for a time in boiling water or dilute acid, to completely isolate the separate cells. This may be accomplished, according to Solla (I), in juicy fruits by oxalic acid or tartaric acid, in potatoes and carrots by acetic acid. The endosperm of *Phytelephas* is, moreover, separated into its cells by chlorine-water or caustic potash in a few days, or by hydrochloric acid in two minutes.

In general, however, more energetic reagents are necessary for the isolation of the separate cells, or at least produce that result more quickly and certainly.

9. The following macerating agents are especially adapted to more general use.

1. *Schulze's Maceration Mixture* (HNO_3 and $KClO_3$).— This is still most frequently used for maceration. It is best used by putting small pieces or slivers of the organ to be treated into a test-tube containing about 2 ccm. of ordinary concentrated nitric acid, adding some crystals of potassium chlorate, and then warming the test-tube until bubbles are

freely evolved. The reagent is then generally allowed to act for a few minutes until the pieces are quite white, when the whole contents of the test-tube are poured out into a crystallizing dish filled with water. The pieces of macerated tissue are then placed directly, or, better, after washing in pure water, or also in alcohol, upon a glass slide. Here they can be readily separated, by needles or similar means, into their separate cells.

Often the isolation of the cells of large pieces of tissue which have been treated with the macerating fluid may be effected by shaking the pieces violently in a glass half full of water.

It should be remarked that the heating of this macerating mixture should preferably be carried on under a hood, or at least not in the neighborhood of a microscope, on account of the evolution of injurious gases.

2. *Chromic Acid* (CrO_3).—Chromic acid is especially useful for the isolation of the cells of sections. These are placed in a concentrated aqueous solution of the acid, and, after it has acted for half a minute to five minutes, are washed in a large quantity of water. The sections are usually then readily separated into their cells. Chromic acid attacks the cell-membranes much more strongly, when acting longer, than does Schulze's mixture, which is in general preferable, although its manipulation requires somewhat more care.

3. *Caustic Potash.*—Caustic potash is useful, especially with delicate tissues, such as the roots of *Taraxacum officinale*. The tissues should be boiled for a few minutes in a solution containing about 50% of potassium hydrate and then placed in water. The cells are then readily isolated by teasing. Dilute caustic potash is also recommended by Solla (I) for the isolation of cork cells.

4. *Glycerine and Sulphuric Acid.*—According to the method proposed by A. Fischer (IV, p. XCVI), it is possible at the same time to recognize starch in the isolated cells. This author places isolated vascular bundles or suitable sections in a solution of iodine in glycerine under a

cover-glass, places at the edge of the latter a drop of sulphuric acid, and warms the whole until it steams, for not over a minute. By pressure on the cover-glass a complete isolation of the separate cells may be accomplished, and any starch present, not being dissolved by the sulphuric acid diluted with glycerine, becomes readily visible by being colored by the iodine. In the case of soft parts of plants, leaves and herbaceous stems, this method may do excellent service. But for wood and the like I have not found it suited.

IV. Swelling.

10. Especially to bring out better certain structural relations of membranes and starch-grains, there may sometimes be used the so-called *swelling media*, which produce an increase of volume depending chiefly on increased water-content.

The most used medium is aqueous *caustic potash*, which causes a greater or less swelling according to its concentration. It is, moreover, very well adapted for the study of the swelling phenomena of protein crystalloids.

Concentrated *sulphuric acid* is also a strong swelling medium, and finally quite dissolves membranes consisting of pure cellulose. *Cuprammonia* (cf. § 246) acts similarly.

As a swelling medium for starch-grains *chromic acid* has been frequently recommended.

Finally, Dippel (I) used a solution of *mercuric iodide* in a potassium iodide solution for making clear certain membrane structures. The proper concentration of this solution must be determined for each special case.

V. Clearing.

11. In many cases, where one wishes not so much to study the entire contents of various cells as to determine their general arrangement, the courses of vascular bundles, or the distribution of less soluble cell-contents, as, for example, calcium oxalate crystals or similar bodies, it may be desira-

ble to make considerable masses of tissue as transparent as possible. For this *clearing* of preparations, vegetable anatomists have heretofore used chiefly strong dissolving or disorganizing reagents, like caustic potash, chloral hydrate, etc., which exert a clearing effect chiefly by the solution or swelling of substances that hinder observation.

Clearing media play an important part, especially in all stained preparations; but here the above-named reagents are not applicable, since they would destroy most stains. In this case one must avail himself almost exclusively of the methods for some time employed by zoölogists and anatomists, which consist in placing the preparations in strongly-refractive media like clove-oil, Canada balsam, etc. These clear less by destroying than by equalizing the refractive differences.

We may thus conveniently distinguish between chemical and physical clearing, even though a perfectly sharp line cannot be drawn between the two processes. Indeed the reagents that are primarily chemically active often have besides a clearing effect due to their higher refractive indices. Yet here there are always two essentially different methods involved, and a separate consideration of them seems to me justified. I will, however, remark that the physical clearing methods are in no way limited to stained preparations, and can be used with the best results, especially in investigations with polarized light.

A. CHEMICAL CLEARING METHODS.

12. Formerly caustic potash was almost wholly used for the chemical clearing of preparations. More recently various other clearing media, especially phenol, chloral hydrate, and eau de Javelle, have been recommended, and in most cases decidedly deserve preference. According to the object of the investigation, one may use sometimes one and again another medium with the best results. As to the manner of using these reagents, the following may be said.

1. *Potassium Hydrate* (KOH) is used mostly in aqueous

solution and in various degrees of concentration. Besides, solutions of potassium hydrate in alcohol or in various mixtures of alcohol and water are recommended for clearing. For complete clearing several hours are often necessary, sometimes even several days, though it can often be hastened by warming. After the removal of the solution and superficial washing with water, the free potassium is best neutralized with dilute hydrochloric or acetic acid. If the preparations are then too opaque, they can be again treated with caustic potash or made more transparent with ammonia.

The well-washed preparations can usually be preserved for a time in glycerine-gelatine; but after a few years they usually become dark and often cloudy also.

For staining preparations treated with caustic potash Errera (IV) recommends *canarin*, which is not decomposed by that reagent.

2. *Phenol* (C_6H_5OH).—The best for clearing is a solution of crystallized phenol, which contains only water enough to keep it fluid at ordinary temperatures. This penetrates cut parts of plants relatively fast, and usually makes them fully transparent in a short time. The clearing may be markedly hastened by heating the objects in the phenol solution to boiling; and at the same time the air is completely expelled from the intercellular systems.

Objects cleared with phenol can, according to my experience, be well preserved in Vosseler's turpentine (cf. § 27).

3. *Chloral Hydrate* ($CCl_3.CH(OH)_2$) has heretofore been used commonly for clearing in a concentrated aqueous solution. This can be used as well with fresh as with alcoholic material. To hasten the extraction of chlorophyll one may use successfully a concentrated alcoholic solution of chloral hydrate. The reaction may also be hastened by warming.

4. *Eau de Javelle.*—A solution of potassium hypochlorite (KClO) is known in pharmacy as Eau de Javelle (Javelle water). This can at any time be obtained ready for use from an apothecary; but it may also be prepared by adding to a concentrated aqueous solution of chloride of lime a solution of potassium oxalate, as long as a precipitate is

formed. The solution, filtered from the precipitate, can be diluted with water before use (cf. Strasburger, I, 632).

Eau de Javelle acts like chloral hydrate, but has the advantage that it destroys chlorophyll much more quickly. It may also be used to decolorize dried parts of plants.

B. PHYSICAL CLEARING METHODS.

13. Physical clearing will evidently be the more complete as the refractive index of the enclosing fluid approaches that of the cell-membrame or of those constituents of the cell-contents which are concerned. Thus in many cases *glycerine* must exercise a certain clearing effect, since the refractive index of pure glycerine (1.46) is markedly higher than that of water (1.33). A much more complete clearing is, however, obtained by various ethereal oils, balsams, and resins. Among these *Canada balsam* plays the chief part at present, and we will therefore first describe in detail the transfer to this medium. This requires, when one has the object in water, a series of manipulations.

Thus, as Canada balsam is quite insoluble in water, a complete *dehydration* of the object is first necessary. Since this is commonly accomplished with alcohol, with which Canada balsam does not mix, the replacement of the alcohol by a fluid which will mix with Canada balsam, xylol, clove-oil, or the like, is required.

In this method, which may be termed the normal or ordinary method of transfer from water to Canada balsam, three distinct manipulations are to be distinguished: dehydration, replacement of alcohol, and transfer to the enclosing medium. Whenever in the following pages transfer to or clearing in Canada balsam is mentioned, the use of this method, which may perhaps seem somewhat complicated to beginners, is understood. After the description of the details of this method, we shall see that the same object can be accomplished by various other methods, whose use is necessary in some cases; for example, when the nature of the stain forbids the treatment of the preparations with alcohol.

Finally, under this head, the use of some other strongly refractive media will be described.

1. **The Ordinary Method of Transfer from Water to Canada Balsam.**

(a) *Dehydration.*

14. Dehydration by alcohol commonly does not present the least difficulty. In case of microtome sections it is sufficient to cover them with alcohol and then let the alcohol flow off; free-hand sections are best placed in small dishes or cups with alcohol, and left for a longer or shorter time, according to their thickness.

But there often results from direct transfer from water to alcohol the shrinking of the cells or their collapse from the too rapid withdrawal of their water. Several methods have been employed to prevent this collapse of the cells, whose essential feature lies in the very gradual replacement of the water by the alcohol.

This can be effected by placing the preparations in turn in different mixtures of water and alcohol, each of which exceeds the previous one in its proportion of alcohol. For instance, one may prevent collapse by placing the preparations first in 10% alcohol and then in order in 30%, 50%, 70%, 90%, and finally in absolute alcohol. The time between the transfers must depend upon the thickness of the tissues. With delicate objects, as, for example, unicellular algæ, intervals of a few minutes each are sufficient.

In the case of filamentous algæ the transfer can be much simplified by binding them together with a thread.

15. Gradual dehydration can be accomplished by a method devised by J. af Klercker, which consists in allowing absolute alcohol to flow slowly into 10% alcohol through a fine capillary tube.

16. The dehydrating vessel* recommended by Fr. E. Schulze (I) brings about the gradual replacement of water

---

* This may be obtained of Warmbrunn, Quilitz & Co., Berlin, C., Rosenthalerstr., 40, at the price of Mk. 2.75 (67 cents).

by osmotic action and is especially adapted for small objects. As is shown in the accompanying Fig. 4, in which the bell-shaped cover that closes the vessel is not shown, this consists chiefly of two cylinders, broadened at the top and placed one within the other, their lower ends being closed by a membrane which permits osmotic exchange between water and alcohol. Schulze recommends for this purpose a thin writing-paper known as " Postverdruss,"* which is glued to the ground lower edge of the cylinder.

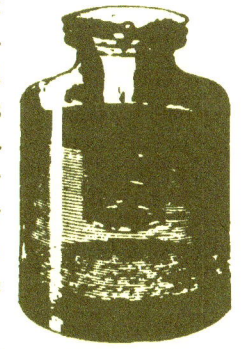

FIG. 4.—Dehydrating vessel. After F. E. Schulze.

In the inner cylinder are placed the objects to be dehydrated in very dilute, about 10%, alcohol; in the outer cylinder is placed a small quantity of stronger, about 50%, alcohol; and in the vessel containing the cylinders is absolute alcohol, which is kept water-free by a layer of anhydrous copper sulphate on the bottom of the vessel. For complete dehydration a period of twenty-four hours is always sufficient. Further, the rapidity of the osmotic interchange may be largely regulated by changes of the differences in level between the different fluids. With less sensitive objects one may find one cylinder sufficient, and then the dehydration can be accomplished in a few hours.

17. According to the method proposed by Overton (I, 12), dehydration may be conducted by placing the objects first in 10% glycerine. In this, objects which have been fixed never suffer collapse; and living ones may first be killed by osmic-acid fumes (cf. § 308). The preparations are then left exposed to the air without a cover-glass, but protected from dust by a bell-jar. The solution of glycerine is thus so concentrated by the evaporation of the water that finally a transfer to alcohol is possible without any collapse.

* [The American papers manufactured under the name of "parchment-paper" and the finer grades of the so-called "Overland paper" serve the purpose well. Best of all is true parchment; and chamois-skin has also been recommended.]

(*b*) *The Replacement of Alcohol.*

18. In earlier years *clove-oil* was almost exclusively used for the replacement of alcohol, and it is in many cases very well adapted for that use on account of its complete miscibility with alcohol. For microtome sections fastened on the slide it is sufficient to place a few drops of clove-oil on the slide after the removal of the alcohol. Thicker sections, especially free-hand sections, are best placed in a vessel with clove-oil, in which they are left until they are completely transparent and no longer appear white on a dark ground.

But clove-oil has the disadvantage of washing out many stains, and has been at present wholly given up by many microscopists, on account of its oxidizing characteristics. How far the other ethereal oils proposed as substitutes for it—oil of origanum, oil of lavender, and others—are free from these disadvantages remains to be discovered.

19. But in any event we have in *xylol* a reagent which can very well replace clove-oil. With microtome sections I use it now exclusively except where I wish to utilize the differentiating effect of clove-oil, as, for instance, in Gram's method (§ 321).

Xylol has only the disadvantage that it mixes with alcohol less readily and requires a more complete dehydration than clove-oil. In consequence of this one easily finds milky cloudings, and with thicker sections would better use, between the alcohol and the xylol, a mixture of three volumes of xylol and one volume of alcohol. For microtome sections it is sufficient in most cases, on the other hand, to cover them with the ordinary so-called absolute alcohol (98%) and then to add xylol. The beginner will do well before the final enclosure in Canada balsam to always examine the preparations on a dark ground. If they appear white and opaque, alcohol should be added again, and then xylol again, until the preparations have become completely transparent.

I will remark here that in this case, and in general when it is desired to bring somewhat large quantities of fluid upon the slide, one may use with satisfaction the bottle shown in Fig. 5, whose hollow stopper ends in a glass tube, while the upper end is closed by a rubber cap. By compressing the rubber cap and then allowing it to expand, fluid is drawn into the stopper, and may then be pressed out in suitable quantities by renewed pressure on the rubber cap. With reagents which are to be used only in drops, the bottle figured in Fig. 6, whose stopper is drawn out simply into a glass rod, may be used.

Fig. 5.—Glass bottle with pipette. After W. Behrens (I).

Fig. 6.—Glass bottle with glass rod on the stopper. After W. Behrens (I).

20. In order to prevent the collapse of delicate objects when brought into clove-oil or xylol we may use the methods proposed by Overton (I, 12).

I. If an object is to be brought into clove-oil or other ethereal oil, it is taken from the alcohol and placed in a small dish containing a 10% solution of the oil in alcohol. This dish is then placed in a somewhat larger one or in a suitable exsiccator, whose bottom is covered with solid calcium chloride. The alcohol is then gradually absorbed by the chloride, and the object becomes at last completely saturated with oil. To prevent longer action of the alcohol, one may transfer the objects from alcohol to water-free chloroform and thence to a 10% solution of clove-oil in chloroform, from which, as in the last-described method, the chloroform may be absorbed by calcium chloride.

II. For transfer to *xylol* the objects are put in a dish with a 10% solution of xylol in alcohol and placed in a exsiccator on whose bottom is pure xylol. Such an adjustment then takes place between the two fluids by diffusion, that the objects finally lie in nearly pure xylol.

21. The transfer of very small objects from alcohol to

xylol may also be accomplished by means of the *settling-cylinder** recommended by Fr. E. Schulze (I), which makes possible at the same time the transfer from xylol into Canada balsam. In this vessel, whose construction is evident from the accompanying Fig. 7 without further explanation, are placed three different fluids in layers above one another. Below is xylol-Canada balsam, next xylol, and finally alcohol. In the latter are placed the previously dehydrated objects. If they are pretty small they sink so gradually to the bottom that they come into the Canada balsam without collapsing. By means of the cock on the side of the cylinder the xylol and alcohol may be drawn off, and the preparations may then be removed directly to balsam on the slide.

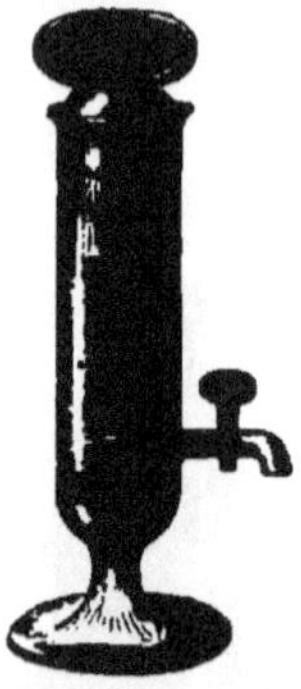

FIG. 7. — Settling-cylinder. After Fr. E. Schulze.

(c) *Transfer into the Enclosing Medium.*

22. For enclosing zoological or botanical preparations in Canada balsam one ordinarily uses fluid Canada balsam, prepared by dissolving this resin in chloroform, xylol, or some similar solvent. The solution in xylol is especially to be recommended. This may be poured into a wide-mouthed glass (cf. Fig. 8), whose cover fits over the mouth outside, and is so high that there is room for a small glass rod in the closed vessel.†

No collapse occurs in transferring an object from clove-oil or xylol to this fluid Canada balsam, as a rule. For very delicate objects the ordinary xylol-balsam may well be diluted with xylol, and then the latter may be allowed to gradually evaporate.

FIG. 8.—Canada-balsam glass, about ⅓ nat. size.

---

* This may be obtained of Warmbrunn, Quilitz & Co., Berlin, C., Rosenthalerstr., 40, at the price of Mk. 3.25 (80 cents).

† Such glasses among others may be obtained at 60 pf. (15 cents) each of Dr. G. Grübler (Leipzig, Bayerische Str., 12).

That collapsing may also be prevented by the use of Schulze's settling-cylinder has already been noticed (cf. § 21).

2. The Transfer from Water to Canada Balsam Without the Use of Alcohol.

(a) *By Drying*.

23. A transfer from water to Canada balsam can be accomplished without the use of alcohol by simply letting the preparations dry in the air, then covering them with xylol, which usually penetrates quite dry sections rapidly, and then enclosing them in xylol-Canada balsam. Naturally this method is applicable only to such preparations as suffer no collapse from drying, especially to very thin microtome sections.

(b) *With Aniline* ($C_6H_5NH_2$).

24. Since about 4% of water is soluble in aniline, the latter can be used for dehydration. The preparations are brought directly from water into aniline, and may then be mounted in Canada balsam. The aniline may be dehydrated by solid potassium hydroxide (KHO), which is wholly insoluble in it (cf. Suchanek I).

(c) *With Phenol* (C_6H_5OH).

25. If sections are transferred from water to phenol which has been melted by warming in a paraffine oven (cf. § 47) or by the addition of a little water, they are cleared in a short time and sufficiently dehydrated to be transferred directly to clove-oil or xylol.

To prevent the collapsing of very delicate objects, the method proposed by Klebahn (I, 419) may be used. The fixed and stained objects are first placed in dilute glycerine, which is allowed to concentrate in the air. Then phenol is added, and clove-oil or creosote is gradually mixed with it, when the objects may be directly transferred to Canada balsam. Klebahn used these methods especially in the study of the germinating spores of Desmids and carried on the various manipulations on slides with hollows ground in them.

3. The Use of Other Strongly Refractive Mounting Media.

(a) *Dammar Lac.*

26. Dammar lac is best dissolved in equal parts of benzol and oil of turpentine. Its use is the same as that of Canada balsam, from which it differs in its somewhat lower refractive index. Thus differences of structure that depend on differences in refractive power often become more conspicuous. The glasses described in § 22 may, of course, be used for this medium. It appears to have been employed little in botanical microscopy.

(b) *Venetian Turpentine.*

27. This mounting medium proposed by Vosseler (II) is prepared by thinning the resin obtained from the apothecary under this name with an equal volume of alcohol and then warming it on a water-bath, shaking it energetically and finally filtering. The filtrate is then somewhat thickened on the water-bath.

The fluid so obtained has the advantage of mixing without cloudiness with 90% alcohol, and thus it makes preliminary clearing with clove-oil unnecessary in cases of incomplete dehydration. Besides, even with delicate objects, like *Spirogyra*, transfer from alcohol to Venetian turpentine much less often causes collapse than does transfer to clove-oil, Canada balsam, and the like. One may avoid the crumpling of very delicate objects by placing them first in a mixture of 10 parts turpentine and 100 parts alcohol, and then permitting a gradual concentration of the turpentine over anhydrous calcium chloride, according to the method of Pfeiffer (I, 30). If small dishes are used for this purpose, they may be provided with ridges of paraffine, to prevent the turpentine from rising on their sides, by simply dipping them to the proper depth in melted paraffine.

The refractive index of Venetian turpentine lies between

that of glycerine and that of dammar lac, so that cell-membranes and starch grains stand out pretty sharply in it.

A disadvantage of this medium lies in the fact that it becomes solid very slowly. In order to secure a firm attachment of the cover-glass to the slide, which is often very desirable in studies with immersion lenses, one may apply a heated metal wire to the edge of the cover-glass, as Vosseler (II, 297) has done. The same object may also be attained, according to Pfeiffer (I), by encircling the cover with Canada balsam.

Various stained tissues, especially carmine, hæmatoxylin, and saffranin preparations, may be excellently preserved in Venetian turpentine, according to Vosseler. In my own experience, however, acid fuchsin seems to be more poorly preserved in it than in Canada balsam.

VI. Staining of Living Tissues.

28. As has been shown especially by the researches of Pfeffer (II), it is possible in very many cases to cause living plants and parts of plants to take up certain coloring matters. This so-called *live staining* is not only of great importance for the study of the transportation of material within the vegetable organism, but has also led to some interesting results concerning the morphology of the cell, and should certainly be capable of still wider application.

For the success of live staining it is of primary importance that the staining solution used should exercise no injurious effect upon the objects concerned. Since the aniline colors generally act as poisons on plant-cells, it is necessary to use them in very dilute condition, when they affect the cell very little or, in general, not injuriously. An evident staining can, it is clear, only take place when the stain is stored up by certain constituents of the cell. This is generally the case when the osmotic balance between the cell fluid and the surrounding staining solution is constantly destroyed by a chemical metamorphosis of the stain taken up. But to make possible in this way the storage of large quantities of staining material, it is also necessary to furnish the objects a large quantity of the staining solution. This staining should therefore be carried on, not on the slide, but in dishes or beakers which will contain at least a half liter of fluid; and not too many of the objects to be stained should be placed in each. Finally, the rapidity of the staining may be hastened by agitating the fluid.

VII. Methods of Fixing and Staining.

29. In the investigation of the various plasmatic constituents of the protoplasm, which are largely colorless and commonly distinguished only by slight differences in refractive power, it is often impossible in difficult cases to reach positive results with living material. But there may be used with the best results the methods of fixing and staining devised chiefly by anatomists and zoologists. These have already led to such important results in the study of the vegetable organism that not the least doubt remains as to their applicability in botanical investigations. Yet, on the other hand, it cannot at all be maintained that the study of living material should now be wholly given up. On the contrary, it should be used, whenever at all possible, for the control and explanation of the results obtained from stained preparations.

30. The purpose of *fixing* is to kill the object in such a way as to preserve its structural relations as completely as possible after the removal of the fixing medium.

It is the object of *staining* to so color certain particular cell-constituents of the fixed preparations, which are to be specially studied, that they shall be sharply differentiated from their surroundings, so that their confusion with other cell-constituents may be prevented.

We possess at present an innumerable lot of fixing and staining methods, and already the most various new or long-known organic and inorganic compounds have been tested with reference to their usefulness as fixing and staining media. While most of these experiments have led to no new conclusions concerning the morphology of the cell, and many methods warmly recommended by their discoverers closely resemble methods long known, yet a thorough trial of all possible salts, acids, and coloring matters should not be omitted. In consequence of the slight insight which we have been able to obtain into the mechanics of staining, it is only possible to discover useful methods in a purely empirical way; and we cannot yet conceive how far newly-discovered stains or new methods may lead toward further conclusions concerning the structure of the protoplasmic constituents of the cell.

31. The result of staining is dependent not only on the nature of the stain used and of its solvent, but also largely on the previous treatment of the object, especially on the fixing medium. Besides, useful results may be obtained after staining by treatment with various solutions of salts, acids, alkalies, and the like, or by the combination of different stains.

We shall become acquainted in the third part of this book with a large number of methods for fixing and staining. But here only the general technique of fixing and staining will be described.

A. FIXING.

32. Fixing is generally the more complete the more rapidly the fixing fluid reaches the cells to be fixed; therefore the best results are obtained with solutions as concentrated as possible, so far as they do not cause precipitates or exert any destructive action. Further, small objects are more quickly penetrated by fixing fluids than larger ones, and therefore in difficult cases the smallest possible pieces, even to sections a few cells in thickness, should be placed in the fixing fluid.

It should be especially observed that *cuticle* and *cork* are not easily permeable by most fixing fluids and indeed are quite impermeable by some. One may, therefore, often greatly aid the penetration of fixing fluids by removing suberized membranes as far as possible, or at least by splitting them to furnish points of entrance for the fluids.

From the above it follows that, in objects which have not

been uniformly acted on by the fixing fluids, those parts which lie nearest the cut surfaces deserve, in general, the greatest confidence in their study.

33. If the objects to be fixed are very light, they can be easily attached by strips of filter-paper to the bottom of the vessel containing the fixing fluid. Especially with objects which are with difficulty wetted, it is often useful to inject them with the fixing fluid, which is easily done by the aid of a filter-pump.

It is not possible to give general statements as to the *time* necessary for complete fixing. This depends, aside from the size of the object, primarily upon the character of the fixing fluid used, and sufficiently exact statements on this point will be given in the description of the different methods.

So, too, the quantity of fluid to be used varies much. In general, one needs relatively little of the energetic fluids, like those containing sublimate, for example; while, especially where potassium bichromate is employed, the use of large quantities of fluid is recommended.

34. For the fixing of objects which easily turn black Overton (I, 9) recommends alcohol containing sulphurous acid. He prepares this by adding to ½ gram of sodium sulphite (Na_2SO_3) a few ccm. of 80% sulphuric acid and conducting the fumes of sulphurous acid which arise, directly into 100 grams of alcohol. Picric acid dissolved in water or in 30 to 50% alcohol may be combined with sulphurous acid in the same way. In preparations treated with the fluids named, the finest protoplasmic differentiations are preserved, and staining with hæmotoxylin or carmine succeeds finely.

B. REMOVAL OF FIXING FLUIDS.

35. It is necessary, in general, before staining to remove completely the fixing fluid used. The fluid to be used for this *washing* of the preparation depends upon the character of the fixing medium. If this is readily soluble in water, it is best to use running water, and for this purpose the *draining-boxes* recommended by Steinach (I) are well adapted, as

also for other uses. These contain, as Fig. 9 shows in section, a glass drainer which rests on three small glass feet and has in its bottom many perforations which become wider downward. The outer box serves to enclose the objects air-tight, for other uses.*

FIG. 9.—Steinach's draining-box in vertical section. After Steinach (1).

If many objects are to be washed at the same time in running water, a useful aid is the washing device which was set up two years ago (1890) in this botanical institute, and which has proved very satisfactory.

This apparatus, whose construction is readily understood from the accompanying Fig. 10, consists essentially of a

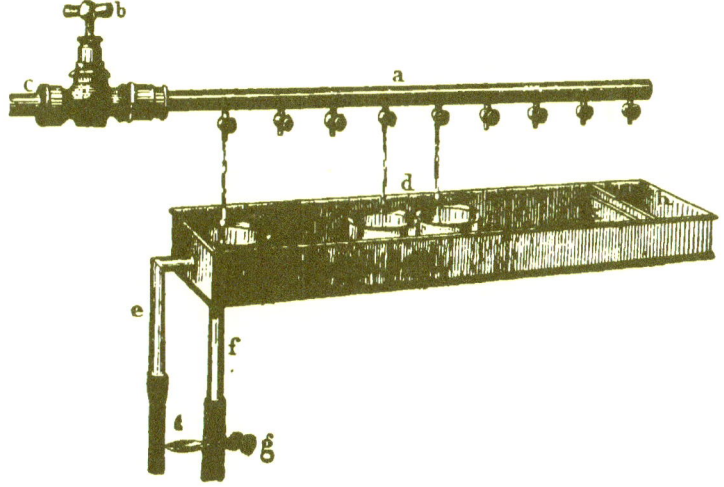

FIG. 10.—Washing apparatus.

brass tube a provided with nine small cocks and the zinc vessel d for the reception of the objects to be washed. But since the small cocks cannot sustain the full pressure of the water-pipes, the complete shutting off and the approximate regulation of the water pressure may be accomplished by means of the large cock b, which, by means of a T-tube, as

---

.* These boxes can be obtained of R. Siebert, Wien VIII, Alserstr. 19, and latterly also of Dr. Grübler. The latter furnishes the glass drainers alone at Mk. 1.25 (31 cents) each.

at *c*, can easily be fitted laterally to any water-faucet. In the zinc vessel the larger space *d* serves to hold the glass drainers. If the water is to run rapidly from it, the pinch-cock *g* is opened, so that the water runs out through the tube *f* that communicates with the bottom of the vessel. If the water is to run from it more slowly, the pinch-cock *g* is closed, and the water can then escape only through the tube *e*, whose mouth is 15 mm. above the bottom of the vessel, so that the water stands 15 mm. deep on its bottom. For the use of the space *h*, see § 39.

C. STAINING.

36. If large objects have been fixed, it is usual, after washing, to cut them into sections and to stain these. In many cases, however, very good results are obtained by staining the objects directly after washing and then preparing sections from them. In case of this so-called staining in mass ("Stückfärbung"), the slight permeability of the cuticle must again be noted, and the penetration of the stain must be aided by its entire removal or by slitting it.

In many cases of mass staining it may be useful to obtain on one and the same section all the different grades of staining side by side, the parts nearest to the surfaces being most deeply stained, while the intensity of the stain gradually decreases as the distance from the surface increases.

If a large number of objects are to be stained at the same time, one may conveniently use the glass drainer described in § 35. Especially with smaller parts or sections of plants, one often finds the glass vessel shown in Fig. 11 useful. Its cover has a groove corresponding to the edge of the dish.*

Fig. 11.—Glass dish with cover, in median section.

If concentrated staining solutions are used in this case, it is often difficult to recognize the separate objects clearly. One may then find

---

* [These so-called "Stender dishes," as well as the other items of glassware described in these pages, may now be obtained of the leading American dealers in microscopical supplies.]

convenient a little apparatus to which Ranvier has given (I, 66) the name *photophore* (light-bearer), but which may better be called, with Obersteiner (I, 55), *section-finder*. It consists of a wooden box, about 5 cm. high and 10 to 12 cm. long and wide, whose front side is wanting, while the top is replaced by a plate of clear glass. There is placed in this box a small mirror so inclined toward the front that it forms an angle of 25° to 30° with the bottom. If this mirror is now turned toward a brightly-lighted window, it will, plainly, reflect the light against the plate of glass forming the top of the box and against the dish of staining fluid placed upon it. One can easily prepare such an apparatus.

The apparatus described by Eternod (I, 41), shown in Fig.

FIG. 12.—Eternod's apparatus.

12, is also very convenient. In this the front part of the glass plate *c*, which is lighted by the mirror *b*, serves as a section-finder, while beneath the hinder part of the glass plate *d-d* is a strip of paper divided into variously colored portions. The outlines, prepared with a diamond, at *g* serve, as is plain without further explanation, for the centering of preparations on the slide, and the small turn-table *a*, for preparing cement rings, etc.

37. *Microtome sections* which are to be stained after cutting are usually fastened to the slide (see §§ 50–52). If it is desired to place a large number of sections in the same staining fluid at the same time, this may be readily accomplished with the aid of a number of crystallizing dishes placed inside of each other, as shown in Fig. 13. The space between two dishes is then filled with the stain-

ing fluid, and the slides are placed in it so that the sides bearing the sections are turned outward. Any injury to or scraping off of the sections is thus prevented. It is best not to fasten the dishes together, but simply to place them inside of each other and to load the inner one with shot or other weight, to prevent their being floated by the fluid in the outer dish. In order to distinguish the different objects in the apparatus from each other, we may use crystallizing

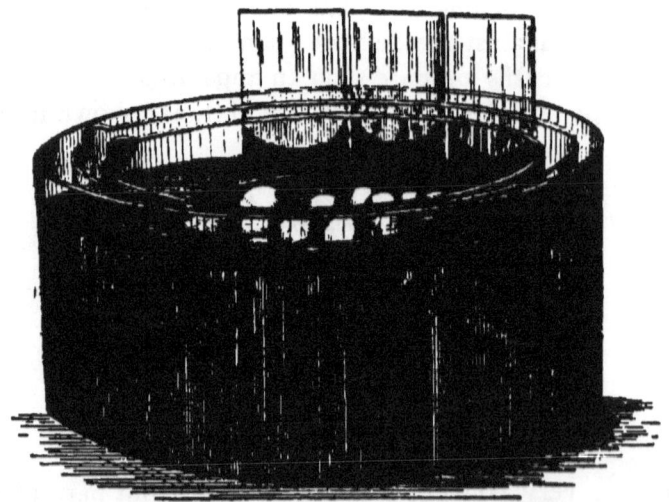

Fig. 13.—Apparatus for staining microtome sections.

dishes provided with lips for pouring. If then the apparatus is always filled in regular sequence in the same direction from this lip, it is only necessary to note the numerical positions of the separate preparations.

38. For further details the reader is referred to the description of individual methods. Here it may be said that preparations are rarely to be studied in the condition in which they are taken from the staining fluid. Often very useful effects are obtained by first strongly "overstaining" the preparations and then washing them in water, alcohol, dilute acid, or the like, until only certain parts of the cell contents appear colored. In many cases the treatment of stained preparations with several different fluids is necessary to obtain well-differentiated images.

39. To wash microtome sections in running water the apparatus described and figured in § 35 may be conveniently used by placing the slides in an oblique position in the space h of the zinc vessel, so that the sides bearing the sections are directed downward. To provide for the constant escape of water from the bottom of this space, the zinc strip i is perforated like a sieve near its lower edge, while the somewhat lower strip, k, is unperforated, but serves to prevent the too rapid fall of the fluid in the space h.

D. FIXING AND STAINING MICROSCOPICALLY SMALL OBJECTS.

40. The fixing and staining of microscopically small objects offers certain technical difficulties, especially if one has but a limited quantity of material. These are best overcome by the methods devised by Overton (I, 13), which can be variously modified to meet the peculiarities of the objects concerned. Thus it seems to me more cenvenient to carry on the manipulations to be described on the slide rather than on the cover-glass, as Overton recommends, unless for special reasons culture in the hanging drop is necessary.

In the first place, for fixing the objects in a drop of culture fluid an easily removable fixing medium should be used. For this purpose the fumes of iodine are well adapted. They are poured out upon the preparation from a heated test-tube, and are easily driven off again by subsequent warming (2 to 5 minutes in the paraffine bath). For the same purpose the fumes of osmic acid may be used. They are applied by holding the slide, with the objects downward, over the mouth of a bottle containing a dilute solution of osmic acid.*

---

* [Where this inversion of the slide cannot be safely risked, the addition to the culture fluid of a drop of a 1% solution of osmic acid may serve the same purpose.]

After fixing, the objects are transferred to alcohol. This is accomplished by first adding a drop of 10–20% alcohol, which causes no collapse, and then placing the preparation in a close chamber saturated with alcohol vapor, to bring about a gradual concentration of the alcohol. For this purpose a flat crystallizing dish may be used, its upper edge being ground so as to be hermetically sealed by a greased glass plate. Its bottom is covered with absolute alcohol and a stand to hold the preparations is placed in it. The latter may be made by simply bending down the ends of some strips of sheet zinc. According to Overton, the cover-glass with the objects is placed, with its wet side upward, on a piece of elder pith, about 3 mm. high and of a diameter less than that of the cover-glass, which, in its turn, rests on an ordinary slide. In this apparatus, which must be protected from sudden changes of temperature and especially from direct insolation, the 20% alcohol on the cover-glass becomes almost absolute alcohol in a few hours, by diffusion through the air. When this is accomplished, a drop of a dilute solution of celloidin is placed on the cover-glass and evenly spread over its upper surface by tipping it backward and forward. It is of advantage to make the celloidin film as thin as possible, since thicker films both separate more easily and render the subsequent manipulations much more difficult. Therefore pretty thin solutions of celloidin must be used. A suitable one may be prepared by diluting the ordinary officinal solution of celloidin with ten times its bulk of a mixture of equal parts of alcohol and ether.*

As soon as the celloidin no longer flows evidently, the whole cover-glass (or slide) is placed in 80% alcohol, wet side up. Here the celloidin film becomes so hard in a few minutes that the objects can be placed in suitable staining fluids without being washed away. A long-continued action of alcohol of more than 90% is to be avoided, as it dissolves

---

* [The thinnest of the celloidin solutions recommended for use in imbedding (cf. § 49a) may be diluted with twice its own bulk of the alcohol-ether mixture for this purpose.]

the celloidin film. Therefore Overton dehydrates with 80 to 85 % alcohol and uses for the transfer to Canada balsam, creosote, which will mix with even 70% alcohol. From the creosote the preparations are either placed directly in Canada balsam or are passed first through xylol. Aniline may be used for the same purpose by passing the objects from 90% alcohol to aniline, then to a mixture of equal parts of aniline and xylol, and then to xylol (cf. § 24).

It should be noted also that many stains, as, e.g., Gentian violet, stain the celloidin film strongly and are, therefore, not to be used with this method.

VIII. Microtome Technique.

41. While the microtome has been generally used for years by anatomists and zoologists, it has been used by botanists in a comprehensive way only in recent years. But since, so far as I know, no one who has recently taken the trouble to familiarize himself with the technique of the microtome, has denied the great value of microtome methods, it seems superfluous to discuss here in detail their advantages and disadvantages.

I will only remark that the methods described in the following sections are not applicable to very hard objects, especially to woods, while they have given me excellent results with all soft structures and also with most leaves and herbaceous stems or roots.

How far the methods proposed by Vinassa (I and II) for cutting very hard objects, aided by a firmer microtome, specially constructed for this purpose, are capable of general application, I cannot judge for want of personal experience. At all events it would be desirable, where possible, to so modify Vinassa's methods as to preserve the protoplasmic elements in the preliminary preparation of objects.*

---

* [The Providence microtome, devised and sold by Rev. J. D. King, Cottage City, Mass., is especially constructed for cutting hard objects and is said to be well adapted to its purpose, but I am not able to speak from experience of it.]

42. I refrain from describing here in detail the various microtomes and their manipulation, and merely remark that I have obtained the best results with a relatively small microtome by Schanze, namely, sections of 1 micron and even of fractions of a micron in thickness.

In this instrument, as may be seen in Fig. 14, the movement of the knife, as well as the raising of the object, is accomplished by the aid of screws. To obtain very thin

FIG. 14.—Microtome by Schanze.

sections, one turns the disk which is connected with the object-raising screw by a system of cogs. This shows directly 1 μ of thickness and permits the estimation of fractions of that thickness.*

I have also worked for a long time with a more elaborate microtome by Aug. Becker (Göttingen) which was very exactly constructed.

[I have used with great satisfaction for serial sections of objects imbedded in paraffine the Minot microtome. This has the knife fixed, while the object is moved vertically past its edge, being pushed forward by an amount equal to the desired thickness of a section, at each descent, by an automatic device. The operation of the instrument consists in

---

* This microtome was constructed from the specifications of Prof. Altmann and may be had of the mechanician M. Schanze (Leipzig, Brüderstr. 63).

the rotation of the balance-wheel by hand power or by a water motor, and its work is thus more uniform and more rapid than that of the sledge microtomes of the Schanze and Thoma types. It is not suited for cutting objects in celloidin.*]

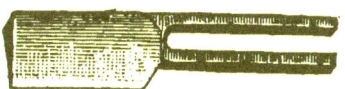

Fig. 15.—Microtome knife. After Henking.

Of the many microtome knives used I have found most suitable that recommended by Henking (1) with a very short edge † (cf. Fig. 15).

I must particularly describe the imbedding of objects to be cut and the manipulation of microtome sections, especially their attachment to the slides. But it cannot be my duty to bring together the very numerous methods recommended by various authors. It will be better for me to confine myself to the careful description of a few methods whose trustworthiness I have had opportunity to prove. Therefore I will particularly describe, among the various modes of imbedding, only the paraffine method, which is by far the best adapted to vegetable objects.

1. Imbedding in Paraffine.

43. For imbedding in paraffine, objects stained in mass or unstained objects may be used. If one is concerned with protoplasmic structures, these must, of course, be carefully fixed and the fixing medium must be washed out before imbedding.

The size of the pieces to be imbedded depends naturally upon the nature of the object. In general it is advantageous to use as small pieces as possible, for, on one hand, these are more easily penetrated by the various fluids, and, on the

---

* [This microtome is sold by the Franklin Educational Co., Hamilton Pl., Boston, at $60, with knife.]

† These are to be obtained of W. Walb (Heidelberg, Hauptstr. 5) under the name of "Henking's microtome knife," at the price of Mk. 4.50 ($1.10) each.

other hand, it is the easier to obtain very thin sections the smaller the surface to be cut is.

44. It is also in no respect unimportant what sort of paraffine is used. I have used in most cases a paraffine recommended by Altmann which melts at 58° to 60° C.* and is obtained from the drug-store of Franz Wittig (Leipzig). But when the size of the sections is more important than their extreme thinness, as may be the case where one wishes a general view, paraffine to which has been added more or less of the superheated paraffine recommended by Count Spee is more useful. Objects imbedded in this have the advantage of rolling up much less easily during cutting.

This superheated paraffine may be prepared by heating ordinary paraffine in an open dish for one to six hours until it has assumed a brownish-yellow color like that of yellow wax, with the evolution of disagreeable white fumes, a slight reduction of its volume, and the elevation of its melting point. Recently such superheated paraffine can be obtained directly from Dr. G. Grübler † and others.

45. In all cases a complete *dehydration* must precede the transfer to paraffine. This can ordinarily be accomplished by means of alcohol. Delicate objects are better not transferred directly from water to alcohol, in order to avoid collapse; but one of the methods for dehydration described in §§ 14 to 17 may be used. In general it is sufficient to use between water and alcohol a mixture of equal parts of both fluids, in which the objects are left for an hour or longer. Afterwards they are left in absolute alcohol from six to twenty-four hours according to their size; or even several days, in some cases.

46. From alcohol the objects are passed to a mixture of

---

* [If paraffine of just this melting point cannot be obtained, it may be readily prepared by mixing two paraffines of respectively higher and lower melting points in proper proportions.]

† [Dr. Grübler's stains, mounting media, and other preparations, which are of standard excellence, may now be obtained of Eimer & Amend, Third Avenue, New York, and of the Franklin Educational Co., Boston.]

three parts by volume of xylol to one part of alcohol, in which they remain twelve to twenty-four hours.* From this they are placed in xylol for twelve to twenty-four hours more. Their complete permeation with xylol may be recognized by the preparations becoming transparent.

I may here remark that chloroform, oil of turpentine, and toluol have been used instead of xylol in this transfer from alcohol to paraffine. But I think it doubtful whether these substances offer any advantage over xylol. However, the above-mentioned mixture of alcohol and xylol is decidedly preferable to the clove-oil formerly used between the alcohol and xylol.

47. From xylol the objects are transferred to melted paraffine; but to prevent the collapse which almost always occurs on a direct transfer from xylol to paraffine, it is better to interpolate a solution of paraffine in xylol. I ordinarily proceed in the following manner with the best results. I place in a so-called *bird's trough*† a mixture of xylol and paraffine which is solid, or at least of a thick consistency, at ordinary temperatures; an exact statement of proportions is not important here. On this cold and solid paraffine-xylol mixture I place the objects to be cut and pour over them enough pure xylol to cover them. Then I place the dish uncovered on the top of the paraffine oven about to be described, where the mixture melts gradually so that the objects covered with xylol can sink into it. A further concentration of the xylol-paraffine is brought about by the evaporation of the xylol. After six to twenty-four hours I place the objects in a porcelain dish filled with melted paraffine, which I place in the paraffine oven, while I allow the dish with the xylol-paraffine mixture to cool for future use in the same way.

The objects remain in a dish full of melted paraffine from twelve to twenty-four hours according to their size, but a longer time is seldom necessary.

---

* For all these transfers Steinach's glass drainers are very useful.

† [Any small porcelain dish will serve.]

For a *paraffine oven* I use in Tübingen with the best results an ordinary double-walled drying-oven (cf. Fig. 16) with the mantle filled with liquid paraffine, into which projects a Desaga thermostat. This is so arranged that the temperature of the fluid paraffine is about 63° C.*

FIG. 16.—Paraffine oven.

48. Finally, to enclose the object quite saturated with paraffine in a block of paraffine suitable for cutting, it is convenient to place a drop of glycerine in a watch-glass of about 60 mm. diameter and to rub it over the inner surface of the glass until the glycerine can no longer be seen.†

The watch-glass is then somewhat warmed and filled with melted paraffine. When this has cooled to near its melting-point, which may be known by its hardening at the edges when lightly blown upon, the objects to be enclosed are placed in it and so oriented with a heated needle that suitable blocks of paraffine can be cut from the mass when cool.

In order to prevent crystallization in the cooling paraffine so far as may be, it should be cooled as rapidly as possible. This is best accomplished by placing the watch-glass, as soon as the objects are oriented, upon a large vessel of cold water, where it will readily float if carefully placed upon the surface. For the same reason it is desirable to place the objects near the edge of the watch-glass where the paraffine is thinnest. When the paraffine is quite cold it separates easily from the watch-glass and is then cut into rectangular blocks two or three centimetres long, each of which

---

* The so-called "Naples water-bath" recommended by Paul Mayer (1) is also very convenient. [This bath in various more or less modified forms may be obtained of American dealers and in its best forms is very useful in the work of imbedding and mounting.]

† This serves the purpose of aiding the subsequent separation of the paraffine from the watch-glass.

has near one end an object to be cut. The opposite end is to be put into the object-carrier of the microtome. But first the end containing the object is to be so trimmed down that, while the object is still wholly enclosed in paraffine, the surface to be cut shall be rectangular and as small as possible. In cutting, this rectangle should be so placed that two of its sides are parallel to the edge of the microtome knife, which is placed at right angles to the direction of its motion.

49. It may be observed here that small blocks of paraffine can be easily attached to rectangular blocks of cork, which may then be fixed in the object-carrier of the microtome. It is only necessary to place a few drops of melted paraffine on one face of the cork and then to quickly put the paraffine block upon it and, by running around its edges a heated metal instrument, to cause it to be completely attached to the block. This proceeding is especially convenient if one wishes to change abruptly the direction of sections, as from transverse to longitudinal. If it is desirable to cut off the paraffine block for this purpose, it may be done with a knife heated over a flame, as in this way the crumbling of the paraffine is prevented.

For preserving blocks of paraffine, the boxes used for the so-called Swedish matches are convenient; [or any small pasteboard boxes.]

1a. Imbedding in Celloidin.

[49a. While the above-described imbedding method and medium are unquestionably of the first value in both animal and vegetable histology, the use of celloidin as an imbedding medium has recently become so extended, and the service it renders in many cases where paraffine does not do well, is so good that some account of the manner of its employment should be given here. It has the advantage that no heat is required in the process of imbedding, and that very large sections may be cut; and the disadvantage that the knife must be kept wet while cutting, and that the thinnest sections which can be cut from it are relatively thick as compared with those which may be cut from paraffine.

Objects to be imbedded in celloidin must first be thoroughly dehydrated, preferably in Schultze's apparatus (cf. § 16). They are placed in a mixture of equal parts of absolute alcohol and sulphuric ether, and then, after a few hours, in a solution of celloidin in the above mixture, which should contain, according to Busse (I), one part by weight of celloidin to 15 parts of the solvent. After it is thoroughly penetrated by this solution, which will require from a few hours to a few days, according to its size and nature, the object passes to a stronger solution containing one part of celloidin in 11 parts of the solvent; and finally, after well penetrated by this, to a still stronger one with the proportion of one to eight parts. After remaining for a suitable time in the last solution, the object is ready for imbedding. For this purpose, a paper strip may be wound tightly about the end of a small block of suitable size and material, preferably of bass-wood or of vulcanized fibre, so as to form the sides of a box whose bottom is the end of the block. This box is now filled with the thickest solution of celloidin, and in it the object is placed and oriented carefully by means of needles wet with the ether-alcohol mixture. As soon as the solvent has evaporated sufficiently to form a firm film over the surface of the mass, the whole may be immersed in alcohol, where it becomes quite hard in a few hours. Since very strong alcohol dissolves celloidin, it cannot be used; and statements vary widely as to the best strength for this purpose. Busse (II) has found, however, that 85% alcohol gives the best results, both as regards the transparency of the celloidin and the thinness of the sections which may be cut from it.

The paper is removed from about the celloidin mass, after it has hardened, leaving it attached to the block. The mass is now trimmed to present a rectangular upper face, and the block clamped upon the microtome so that the object may be cut in the desired plane.

To cut successful sections from a celloidin block, it is

necessary to set the knife very slightly oblique, and as nearly as possible parallel, to the direction of its motion, so that the celloidin shall be cut with a long drawing stroke. The knife and the top of the block should also be kept wet, during cutting, with alcohol of the strength of that in which the block was hardened. With these precautions excellent sections may be obtained.

Busse (I) recommends the use of photoxylin instead of celloidin, as it gives a more completely transparent imbedding mass. The details of its manipulation are precisely the same as for celloidin. More detailed accounts of the use of celloidin may be found in papers by Eyclesheimer (I) and Koch (I).]

2. The Attachment of Sections.

50. For the purpose of dissolving out the paraffine from microtome sections filled with it, these are commonly attached to the slide. Although recently a large number of methods for accomplishing this have been proposed, I will restrict myself to describing somewhat in detail four of them, each of which seems to possess certain advantages for some cases.

A. ATTACHMENT WITH COLLODION.

In the first of these a solution of about 5% of officinal collodion* is used for attachment. It is conveniently kept in a bottle having a soft brush inserted through its cork. A drop of this solution is first allowed to flow under the sections arranged as desired on a slide, a piece of filter-paper is then laid upon them, and the sections are pressed down upon the slide with the finger or with a paper-knife or similar instrument. Then the sections are painted over with the collodion solution and it is allowed to dry in the air. When this is done, the slide is warmed over a small flame

---

* [Or a mixture of equal parts of a thin solution of collodion and clove-oil.]

until the paraffine melts, and then plunged in xylol to dissolve the paraffine.

After this the sections, if from objects stained in mass, can be at once enclosed in xylol-Canada balsam. But if they are to be stained, they must first be brought into water or alcohol according to the nature of the stain to be used. Since the separation of the sections from the slide has often occurred during this transfer, I now perform it by carrying the preparations from xylol successively into a mixture of three parts xylol and one part alcohol, 90% alcohol and 50% alcohol, leaving them in each fluid two minutes, or as much longer as is necessary.

I use for this purpose vessels with parallel sides, on the bottom of each of which, at one of the short sides, a piece of cork about 1 cm. high has been fastened. The slides are then so placed in these that one end rests on the piece of cork and the side bearing the sections is turned downward.

From 50% alcohol the preparations can be transferred to water or any suitable staining fluid, without fear of the separation of the sections. At least, I have experienced such a result very rarely even in the use of the most complicated staining processes when the above precautions have been observed; and I have not been troubled by any serious staining of the delicate collodion film by any of the more important methods.

[50a. *Celloidin sections*, when arranged on the slide, may be attached to it by placing the whole in a close chamber over ether. The ether vapor quickly dissolves the celloidin sufficiently to cause the sections to adhere firmly to the slide on removal from the chamber. Should any difficulty be experienced, the sections may be arranged on a thin collodion or celloidin film on the slide and then treated as above. After they are attached, they may be stained and mounted as described for paraffine sections (§ 50). Objects stained in mass may be imbedded in cellodin as well as in paraffine.

For mounting in Canada balsam, celloidin sections may be cleared with a mixture of three parts xylol and one part phenol, or with equal parts of phenol and oil of bergamot, or with oil of bergamot alone. The last two are especially recommended. When objects are stained before imbedding, the whole block may be cleared before cutting.]

B. ATTACHMENT WITH AGAR-AGAR.

51. Agar-agar is recommended by Gravis (I) for the attachment of microtome sections. A $\frac{1}{10}\%$ aqueous solution of this substance is warmed for some time after the mixture of the ingredients until quite homogeneous, then filtered through a fine cloth or through glass wool, and finally protected from spoiling by the addition of some pieces of camphor.

A drop of this solution is placed on the carefully cleaned slide, the sections are laid upon this drop, and the whole is warmed until the paraffine becomes soft without wholly melting. Crumpled sections then spread out completely. After the cooling of the slide the superfluous agar-agar is taken up with filter-paper and the rest is allowed to dry completely. After this the paraffine may be removed by xylol, as in the previous method, and the slide may be transferred to alcohol.

This method, which I have recently tried many times, has the advantage that it admits of the use of rolled sections; and even crumpling due to the imbedding may be wholly or largely overcome. I have seen a troublesome staining of the agar-agar only with hæmatoxylin.

A disadvantage of the method, however, consists in the fact that in pure water the solution of agar-agar, and therefore the separation of the sections, often occurs. But one can always treat sections attached with agar-agar with solutions in strong or 50% alcohol, and can usually, with some care, stain them with aqueous solutions.

C. COMBINED AGAR-AGAR-COLLODION METHOD.

The separation in water of sections attached with agar-agar can be prevented by painting over sections attached by the method just described, after they have fully dried, with the above-mentioned collodion solution (cf. § 50) and letting them dry in the air. I have used this method often recently, and can recommend it heartily for difficult cases. It unites the advantages of both methods in that it makes possible the recovery of collapsed sections and permits the use of aqueous stains without fear of separation of the sections.

D. ATTACHMENT WITH ALBUMEN.

52. According to P. Mayer's (I, II) methods, a solution of albumen is used for attaching sections. This is prepared by mixing 50 cc. of the albumen of hens' eggs with 50 cc. of glycerine and 1 gram of sodium salicylate, and filtering the mixture after hard shaking. A small drop of this solution, which, according to Vosseler (I, 457), becomes useless in about six months, is placed on a carefully cleaned slide and is rubbed with the finger or a soft cloth until a barely visible film remains upon the slide. The sections are placed upon this and pressed down upon the slide, a dry brush being held between the finger and the sections. If the slide is now heated over a small flame until the paraffine melts, the sections become so firmly attached by the coagulation of the albumen that the paraffine can be dissolved out with xylol or other solvent without fear of their being washed away. Nor does this occur when they are transferred directly from xylol to alcohol or from alcohol to water. Neither have I observed the staining of the albumen film by any coloring matter; so that this method may be most conveniently used for most cases.

IX. Making Permanent Preparations.

53. One may use very various methods for preserving preparations as long as possible. In nearly all cases preparations enclosed in Canada balsam or some other resin or balsam possess the greatest permanence. But, on account of their high refractive index, which nearly corresponds

with those of cellulose and of most of the contents of the vegetable cell, these substances are hardly to be used except for stained preparations or such as are intended for observation by polarized light. Besides, the transfer of easily collapsible objects to balsam is so complicated that other mounting media are preferable for them. It depends wholly on the nature of the objects to be mounted what mounting medium is best to be used, and it will be necessary, in the second and third parts of this book, to repeatedly indicate what method of preservation is best adapted to the case in hand.

Since the methods of mounting in Canada balsam, Dammar lac and turpentine have been already described in §§ 14 to 27, only the remaining methods, in which glycerine especially plays an important part, need be here brought together.

54. *Glycerine.*—Pure glycerine in various degrees of dilution, or a mixture of this with an acid, was formerly a much esteemed and almost universally used medium. In its use, however, especial care must be taken that the glycerine used is not diluted by too long exposure to the air, since in that case a gradual drying up of the preparation takes place, if the subsequently applied cement ring (cf. § 62) is not airtight. Concentrated glycerine often cannot be used, however, on account of its strong clearing and dehydrating power. In such cases a dilute solution of glycerine with a few drops of acetic acid offers great advantages, but must, as already remarked, be very carefully protected against evaporation. (Cf. especially Dippel, II, 1010.)

55. *Glycerine and Chrome Alum.*—For preserving preparations of *Schizophyceæ* and *Florideæ* in their natural colors, Kirchner uses (I, p. VII) dilute glycerine, to which is added enough chromium-potassium sulphate (chrome alum) to give the fluid a clear bluish color.

56. *Glycerine-gelatine.*—This is of late most used and offers undeniable advantages, in most cases, over the fluid glycerine mixtures. It is conveniently prepared from the recipe recommended by Kaiser, as follows: One part by

weight of gelatine is soaked in six parts of water; seven parts of pure glycerine are then added, and finally a gram of phenol to each 100 grams of the mixture. The whole is then warmed for 10 to 15 minutes with constant stirring, until the fluid is quite clear, and is finally filtered through glass wool or filter-paper. This may evidently be best done with the aid of a hot-water filtering apparatus.

57. Less delicate objects, like sections of wood and the like, may be transferred directly from water to glycerine-gelatine; more delicate preparations should first be brought into glycerine. This may be accomplished, in case of objects which collapse very easily, by placing them in a 10% solution of glycerine, which is then allowed to concentrate gradually by standing in the air.

58. Since the glycerine-gelatine (glycerine jelly) is solid at ordinary temperatures, it must be warmed before use until it becomes fluid; and for this purpose the paraffine bath may be used (§ 47). Or one may prepare small pieces of the jelly, each of suitable size for one preparation, and melt them upon the slides. Such pieces may be readily prepared by allowing a quantity of the jelly to harden upon a plate in a thin layer, which is then cut into blocks.

59. If annoying air-bubbles occur in the preparation enclosed in glycerine-gelatine, they can be easily removed from objects not too delicate by heating the jelly to boiling.

Since preparations in glycerine-gelatine usually shrink pretty strongly when kept for a long time, it is generally advisable to seal them with a cement ring; but it is best, especially with thick sections, to apply this ring after some time, as otherwise the cover-glass is easily broken by the subsequent concentration of the jelly, and it is easier to remove by warming, any air-bubbles that may appear. For demonstration preparations I apply the cement only after a year.

59a. *Chloral-hydrate gelatine* is recommended by Geoffroy (I) as a mounting medium. It is prepared by dissolving 3 to 4 grams of good gelatine in 100 ccm. of a 10% aqueous solution of chloral hydrate, at as low a temperature as possible. The sections are placed directly in this fluid, and, since a thin layer of gelatine is soon formed at the edge of the cover-glass by the evaporation of water, the preparation may be sealed after a short time with maskenlack or an alcoholic solution of sealing-wax. Many stains, such as those with iodine green or carmine, are well preserved in this medium.

60. *Enclosure in Air.*—Ash skeletons, crystals easily soluble in water, and the like may be often best preserved simply dry. To protect them from dust it is also necessary in such cases to cover the objects with a cover-glass. This may be attached to the slide by wax or paraffine around its edges or even with gummed paper.

61. *The observation of crystalline precipitates* and the like is best conducted in the air by ordinary light; while in polarized light the interference colors appear most pure on enclosure in a strongly refractive medium like Canada balsam. Instructive preparations of both kinds may be prepared by placing on the slide a drop of Canada balsam so small that it occupies only a part of the space beneath the cover-glass, leaving a part of the crystals in air. To exclude destructive agencies so far as possible, the edge of the cover-glass may then be surrounded with paraffine or wax.

62. *Sealing Media.*—Of the numerous sealing media proposed by various authors may be mentioned here first the so-called "gold-size," which is well adapted for glycerine and glycerine-gelatine preparations. Since the method of preparing it is quite elaborate, it is best to obtain it ready prepared (e.g., from Dr. G. Grübler, Leipzig).

For glycerine-gelatine preparations Canada balsam, asphalt varnish, and maskenlack N. III are also well adapted. The cover-glass cement containing amber, recommeded by Heydenreich, affords a very trustworthy medium; but it should not be colored with eosin, as was the case with a preparation formerly furnished by Dr. G. Grübler, because this gradually goes over into the glycerine-gelatine and may cause an unpleasant staining of the preparation.

Part Second.

MICROCHEMISTRY.

A. Inorganic Compounds.

1. Oxygen, O_2.

63. For the microchemical recognition of oxygen, the method with *Bacteria* devised by Engelmann (I) may often be used with success. This depends upon the fact that moving Bacteria at once cease their motion if oxygen is withdrawn from them, and immediately resume it on the subsequent renewal of the oxygen supply. Oxygen also affects the direction of motion of Bacteria, since they move toward the fluid which is richest in oxygen.

It is easy to satisfy one's self of this by placing a drop of fluid containing moving Bacteria on a slide, and covering it with a large cover-glass. The oxygen of the fluid is soon exhausted and the motion continues only at the edges of the cover, or around included air-bubbles, which are especially instructive. It may also soon be seen that the Bacteria group themselves in heaps at these places.

The sensitiveness of this reaction, which shows very small quantities of oxygen, is naturally dependent in some degree upon the choice of Bacteria. Those which are obtained by letting split peas decay in water are very useful. After a few days innumerable Bacteria appear, which are commonly called *Bacterium termo*.

It may be added, with reference to the management of the reaction, that it is usually desirable to use *large* cover-glasses, whose edges may be sealed with cacao-butter, wax, or paraffine to prevent the evaporation of the fluid and the access of oxygen.

2. Peroxide of Hydrogen, H_2O_2.

64. For testing living *Spirogyræ* for the presence of peroxide of hydrogen, Bokorny (III) used the two following methods.

The first is based on the fact that peroxide of hydrogen in the presence of iron sulphate at once sets iodine free from potassium iodide, so that any starch or starch-paste present is colored blue. He therefore placed *Spirogyra* cells containing starch in a very dilute solution of ferrous sulphate and potassium iodide, and deduced the absence of peroxide of hydrogen from the failure of the starch-grains to become colored blue. This was emphasized by the intense bluing of the starch in threads which had previously been saturated with the peroxide.

In the second method, Bokorny acted upon the fact that tannin which gives a blue reaction with ferric salts is at once turned blue by ferrous sulphate in the presence of peroxide of hydrogen, while the blue color otherwise appears only after some time in consequence of the gradual oxidation of the ferrous salt in the air. He observed, in agreement with the above, that *Spirogyra* threads containing a tannin that reacts with ferric salts became blue only many hours after being placed in a solution of ferrous sulphate, while the blue color appeared at once in threads saturated with the peroxide.

Pfeffer has (IV, 446) questioned the conclusiveness of these experiments and especially doubted whether the dilute reagents used by Bokorny were really taken up by the living cells. But Bokorny has (I and II) recently made observations which show that the ferrous sulphate is really taken up by the living cells, and the conclusiveness of the second reaction cannot, therefore, be doubted.

65. Pfeffer (IV) was led by more extended observations to the conclusion that peroxide of hydrogen does not occur within the living cell. He showed first that the peroxide may be taken up by living cells without harm and that it often produces in them, even when in very small quantity, plainly visible reactions, which do not otherwise occur in living cells.

Pfeffer used first for these researches plants whose colorless cell-sap is colored by the oxidizing effect of the peroxide, as, e.g., the epidermal cells of the stem and root of seedlings of *Vicia Faba* or the root-hairs of *Trianea bogotensis*. In these the peroxide produces a browning of the cell-sap which is usually followed by the separation of red-brown or almost black granular masses, as is shown in Fig. 17. Here is figured a part of an epidermal cell from the stem of *Vicia Faba*, which has lain five hours in a solution of peroxide of hydrogen, prepared by mixing ten parts of pure water with one part of a commercial peroxide solution already six months old.

FIG. 17.—Part of an epidermal cell of the stem of *Vicia Faba*, five hours after being placed in peroxide of hydrogen solution.

66. Pfeffer also worked with cells which have naturally a colored cell-sap, like the stamen-hairs of *Tradescantia virginica*. In this case the blue cell-sap is wholly bleached by the peroxide or takes a yellow-brown or vinous-yellow color.

Bleaching by the peroxide taken up may also be observed in cells whose protoplasm has been previously colored blue by cyanin. The root-hairs of *Trianea bogotensis* are well adapted for these experiments. In their protoplasm, when in a very dilute solution of cyanin, prepared by warming that dye with water, various blue differential stains were evident in from three to fifteen minutes, and were destroyed by peroxide of hydrogen in less than a minute.

67. It may be remarked that Pfeffer worked with solutions of from .01% to 1% of the peroxide. Since the commercial peroxide always contains some free hydrochloric acid to increase its keeping quality, it must be neutralized with sodium bicarbonate; and Pfeffer adds this in slight excess.

3. Sulphur, S.

68. The sulphur which occurs in various Bacteria in the form of strongly refractive spheres (cf. Fig. 18, 1, *a–c*) is, according to Cohn (I, 178), insoluble in water and hydrochloric acid, but soluble in an excess of absolute alcohol, in hot potash, or in sodium sulphite. Nitric acid and potassium chlorate dissolve them at ordinary temperatures, as does carbon bisulphide; but the entrance of the latter into the cells of the Bacteria must be aided by previously killing them with sulphuric acid or by drying. According to Winogradsky (I, 521), this solubility in carbon bisulphide is not complete, although the insoluble residue in this reagent is always small. According to Bütschli (I, 6), the granules of sulphur are soluble in twenty-four hours in artificial gastric juice, or in a 10% soda solution.

69. Various observations of Winogradsky (I, 518) explain the accumulation of the sulphur granules. According to this author, they are always quite spherical in the living cell and run together on the death of the cells, for example, on heating to 70° C., into large drops which change into beautiful crystals of sulphur. This crystallization takes place best when *Beggiatoa* threads rich in sulphur are placed for about a minute in a concentrated aqueous solution of picric acid

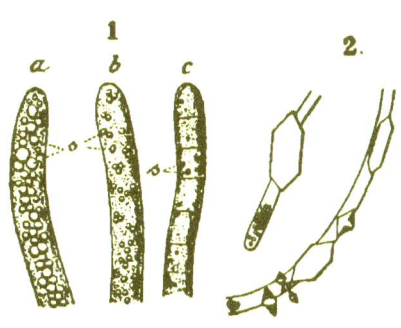

FIG. 18.—1. *Beggiatoa* threads. *a*, very rich in sulphur; *b*, after 24, *c*, after 48 hours' culture in spring-water; *s*, granules of sulphur. 2. The same, 24 hours after treatment with picric acid, which largely converts the sulphur globules into crystals. After Winogradsky.

and then washed in a large quantity of water. On such threads beautifully formed sulphur crystals were found after twenty-four hours, partly monoclinic prisms and partly rhombic octahedra (cf. Fig. 18, 2). It is therefore to be presumed that these sulphur grains consist of the modified form of sulphur which is semi-fluid or oil-like at ordinary temperatures. In fact, the precipitate of sulphur which is formed when dilute hydrochloric acid is added to a solution of calcium pentasulphide shows the same relations, on microscopic examination, as the sulphur granules of the *Beggiatoas*. It is possible, according to Winogradsky, that these, as well as the granules of the precipitated sulphur, gradually pass over into the solid condition, and that, especially in slowly growing threads, all the stages from the fluid to the almost solid condition occur.

70. It should be observed here that Jönsson (I) has seen in a mycelium of *Penicillium*, growing on dilute sulphuric acid, strongly refractive bodies which correspond in many of their reactions with the sulphur granules of the *Beggiatoas*, and consist, according to Jönsson, of a mixture of sulphur and an oil-like substance.

4. Hydrochloric Acid, HCl, and its Salts.

71. For the recognition of hydrochloric acid Schimper (II, 212) found the two following methods especially useful.

I. The addition of *silver nitrate* causes the formation of amorphous silver chloride, but this may be obtained in crystalline form by dissolving the precipitate arising from the addition of silver nitrate in as little ammonia as possible and allowing the fluid to evaporate. Regular crystals of silver chloride are thus formed, consisting chiefly of hexahedra, octahedra, and rhombic dodecahedra, as well as combinations of these (cf. Fig. 19). These crystals gradually become violet-colored in the light; but in the presence of reducing plant-juices they often become very rapidly colored. Formed silver chloride may also be recognized by its ready solubility in potassium cyanide, in sodium hyposulphite, and in a concentrated solution of mercuric nitrate. It is also somewhat soluble in concentrated solutions of the alkaline metals and in concentrated hydrochloric acid; and, accord-

FIG. 19.—Crystals of silver chloride. After Haushofer.

ing to Borodin's method,* silver chloride may be tested with a concentrated solution of silver chloride in concentrated hydrochloric acid or salt solution (sodium chloride).

II. *Thallium sulphate* causes at once, or at least on evaporation, the formation of regular octahedra or variously shaped skeletons of thallium chloride, which may be tested, according to Borodin's method, with a concentrated solution of thallium chloride.

5. Sulphuric Acid, H_2SO_4, and its Salts.

72. For the recognition of sulphuric acid we still lack a completely trustworthy method. The following methods have been used by Schimper (II, 219):

1. *Barium chloride* always causes a precipitate of barium sulphate, but this is rarely crystalline and its positive determination is therefore rarely possible.

2. *Strontium nitrate* causes the formation of small, thick crystals of a mostly roundish-rhombic form, though sometimes sharp and with straight outlines, which are insoluble in water.

3. *Potassium sulphate* often crystallizes out of a solution of ash in water in the form of hexagonal plates, which fall into colorless granules on the addition of barium chloride,

---

* According to Borodin's method (II, 805) a given precipitate soluble in water is tested with a completely saturated solution of the substance that is suspected in it. If the suspicion is correct, the precipitate will not be dissolved, while any other substance, unless some reaction occurs, will be soluble. If, for instance, we have to do with a mixture of asparagin and saltpeter (potassium nitrate) the asparagin crystals will, of course, be insoluble in a concentrated solution of asparagin, but the saltpeter crystals will be dissolved. On the subsequent addition of water, asparagin crystals will be dissolved also. So, as in the above-mentioned case, silver chloride will be insoluble in a concentrated solution of silver chloride in strong hydrochloric acid (or NaCl), while it must dissolve on the addition of more acid (or NaCl solution). In case of substances not too easily soluble, this method renders good service in microchemistry; but great care must be taken in each case that the solution employed is really completely saturated, and that it does not become capable, through changes of temperature, of dissolving more of the substance concerned.

or into heaps of red granules on the addition of platinum chloride.

4. *Sodium* and *potassium sulphates* may often be recognized in the living tissues by means of nickel sulphate. With this they form well crystallized double salts of the composition $NiSO_4 + Na_2SO_4 + 6H_2O$ (or the corresponding K salt); these occur mostly in the form of the monoclinic prism combined with the basal plane, but are pretty easily soluble in water.

6. Nitric Acid, HNO_3, Nitrous Acid, HNO_2, and their Salts.

73. *Diphenylamine* was first recommended by Molisch (I) for the recognition of the nitrates, and he used for fresh sections a solution of from $\frac{1}{100}$ to $\frac{1}{10}$ of a gram of it in 10 ccm. of pure concentrated sulphuric acid, or for dried sections a concentrated solution of it in concentrated sulphuric acid. In the presence of nitrates there occurs immediately after the addition of this reagent a deep-blue coloring which, after a time, disappears or passes into brownish yellow.

This reaction occurs in the same way in the presence of nitrites, and it can therefore be used for the recognition of nitrates only when the absence of nitrous salts is proved. But in fact all investigations on the subject heretofore have led to the conclusion that nitrous salts do not occur within the living plant; and therefore this objection to the applicability of diphenylamine as a reagent for nitrates falls, so far as the microchemical study of the plant is concerned.

It should be remarked that other compounds than nitrates and nitrites give the same reaction, as, for example, manganese peroxide, potassium chromate and chlorate, hydrogen peroxide, ferric oxide and its salts (cf. Frank I and II, and Kreusler I). But these substances appear to be as rare in the plant as nitrites; at all events, plants freed from nitrates never give a blue color with diphenylamine, according to the confirmatory researches of Frank and Schimper (II, 217).

It is more important to note that the reaction may entirely fail, even with large quantities of nitrates, in presence of various substances, as, for example, lignified cell-membranes (cf. Schimper II, 217). It follows, therefore, that the absence of nitrates can never be deduced from a negative result of this test.

74. *Brucin* gives a bright red or reddish-yellow color with nitrates and nitrites, but this gradually disappears. Molisch (VI, 152) uses for microchemical purposes a solution containing .2 gram of brucin in 10 ccm. of concentrated sulphuric acid, but remarks that this reaction is inferior in clearness to the diphenylamine-reaction.

75. According to Arnaud and Padé (I), the alkaloid *cinchonamin* ($C_{19}H_{24}N_2O$), obtained from the bark of *Remijia purdicana*, may be used for the microchemical recognition of nitrates. Its nitrate is almost absolutely insoluble in acidified water and forms beautiful, readily recognizable crystals whose form is, unfortunately, not described by these authors. They immerse fresh sections of the parts to be tested directly in a .4% solution of the chloride of cinchonamin which is slightly acidified with hydrochloric acid. The crystals of nitrate of cinchonamin will then be formed within the cells containing nitrates.

76. *Potassium nitrate* (KNO_3) may also be recognized by covering the sections with a cover-glass, adding alcohol, and then allowing them to dry. The saltpeter then usually crystallizes, chiefly in the form of rhombic plates (cf. Fig. 27, § 130), which stand out sharply, especially in polarized light. Asparagin also forms similar crystals, but these may be easily distinguished from saltpeter crystals by measuring their angles (cf. § 130). Besides, the latter are, of course, easily soluble in a concentrated aqueous solution of asparagin, and are not destroyed by heating. They can also be readily tested with a solution of diphenylamine. Borodin's method (cf. § 71, note) is inapplicable, on the other hand, on account of the ready solubility of potassium nitrate.

7. Phosphoric Acid, H_3PO_4, and its Salts.

77. The following reactions are adapted for the microchemical recognition of phosphoric acid:

1. *Nitric acid and ammonium molybdate.* This reagent, first introduced into microchemistry by Hansen (I, 96), causes the formation of regular crystals which represent chiefly a combination of the octahedron and the cube, and are colored an intense yellow. There is commonly no danger of confusing these with the isomorphic compounds of arsenic acid, so far as the study of vegetable objects is concerned.

It is convenient to use as the reagent a solution which contains 12 ccm. of officinal nitric acid of specific gravity 1.18, to one gram of ammonium molybdate. In the presence of small quantities of acid the precipitate is formed only after slight warming (to 40°–50° C.), and then often only after some time.

The sections to be tested are best burned before the addition of the reagent, since otherwise the reaction may be hindered by the presence of certain organic substances, as, for example, potassium tartrate. Besides, the phosphoric acids combined with the nuclein or otherwise organically united, as, for instance, the phosphoric acids contained in the globoids, are not directly shown by this reagent, but only in the ash (cf. Schimper II, 215). This reagent may be applied directly to the ash prepared by heating upon the cover-glass. Thus is obtained at once with the ashes of sections of not too young stems of *Stapelia picta* a strong reaction, which occurs only after some hours in sections prepared from alcoholic material which contain sphærites of calcium phosphate (cf. § 96).

2. The addition of *magnesium sulphate* and *ammonium chloride* produces with salts of phosphoric acid a crystalline precipitate of ammonio-magnesium phosphate, which is practically insoluble in ammonia and ammonium chloride solutions. These crystals, some of the most characteristic

of which are illustrated after Haushofer (I, 92) in Fig. 20, belong to the rhombic system. A similar salt is also formed by arsenic acid.

A suitable reagent may be obtained by mixing 25 volumes of a concentrated aqueous solution of magnesium sulphate, 2 volumes of concentrated aqueous solution of ammonium chloride, and 15 volumes of water. If there be placed in this solution sections from alcoholic material of the stem of *Stapelia picta*, which have previously been soaked for a time in water to prevent the formation of a precipitate by the alcohol, there appear after a time in the immediate vicinity of the sphærites of calcium phosphate, in consequence of their gradual solution, well formed crystals of ammonio-magnesium phosphate, among which the X-shaped skeleton-crystals appear to be especially characteristic. This reaction may be hastened by warming, but the crystals are then less regularly formed.

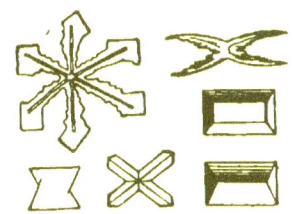

FIG. 20. — Crystals of ammonio-magnesium phosphate. After Haushofer.

For the recognition of phosphoric acid within the tissues, this reaction is, according to Schimper (II, 216), preferable to the previously described one, since it is not interfered with by the presence of organic compounds and is very delicate.

8. Silicic Acid, SiO_2, and the Silicates.

78. Silicic acid occurs in the vegetable kingdom partly in incrustations of cell-membranes and partly in the form of variously-shaped silica masses in the interiors of cells (cf. Kohl's compilation, II, 197).

For the microchemical recognition of silicic acid, one may utilize its peculiarity of not being changed by heating. Its insolubility in all acids except hydrofluoric acid serves to distinguish it from other inorganic substances. In case of some strongly silicified organs it is possible by the combined action of acids and heat to obtain completely coherent siliceous membranes, the so-called silica skeletons. From the membranes of the diatoms, which are peculiarly rich in silicic acid, or from the epidermis of the *Gramineæ* or *Equisetaceæ**, beautiful siliceous skeletons may be obtained by treating them as proposed by Sachs. This method consists in heating the organ or organism on a cover-glass, or on a bit of mica to prevent the residue from adhering, with a drop of concentrated sulphuric acid until the ash remaining after the evaporation of the acid has become quite white.

In case of objects poorer in silicic acid, satisfactory siliceous skeletons cannot usually be obtained by this simple method. It is then commonly better to remove the soluble inorganic substances from the pieces before burning by treatment with hydrochloric or nitric acid. In this way pure white skeletons may be much more easily obtained and may be freed from foreign admixtures by renewed treatment with hydrochloric acid.

79. Besides, siliceous skeletons may be very well prepared wholly in the wet way by the method proposed by Miliarakis (I). The object is first treated in a beaker with concentrated sulphuric acid until it is quite black and then a 20% aqueous solution of chromic acid is added. In this mixture suberized membranes are also wholly destroyed, and only the siliceous skeletons remain behind. They may then be easily isolated, after the addition of water, by decanting, and may be completely cleaned by repeated washing with water and alcohol. The siliceous skeletons of diatoms obtained in this way show, especially when examined in air, the finest structural features of their membranes.

According to Kohl (II, 226) this method is applicable only where considerable quantities of silicic acid are present. This author obtains very delicate siliceous skeletons by burning from parts of plants with a small proportion of silicic acid, which would be completely dissolved by the treatment with chromic sulphuric acid. In other cases the

---

*[Our *Equisetum hiemale* is especially good for this purpose.]

presence of silicic acid can only be recognized in the ash by the sodium silico-fluoride reaction (cf. § 81).

80. To test the skeletons obtained by either of these methods for the presence of silicic acid, *hydrofluoric acid* may be used, in which pure silica-skeletons should dissolve completely.

Kärner (I, 262) recommends for this purpose a dilute aqueous solution of hydrofluoric acid, which, as it attacks glass, must be kept in a bottle of rubber or lead, and must be placed upon the objects to be tested with a platinum wire or a rubber rod. The slide which supports the object must, of course, be protected from the action of the acid, and for this purpose covering with Canada balsam, glue, vaseline, or glycerine has been recommended. But, according to Kärner (I, 262), it is most convenient to cover the slide with a piece of transparent sheet-wax, which must be first somewhat warmed and smoothed by rubbing between the hands. Instead of a cover-glass this author recommends tha use of gelatine paper. He also fastens a bit of the same paper to the objective with Canada balsam, to protect it from the vapor of hydrofluoric acid.

When Kärner (I, 266) allowed hydrofluoric acid to act upon membranes not previously treated with some acid or the like, he usually found only a partial solution of the silicic acid. Whether this was due to the physical action of other constituents of the membrane or to a chemical union, perhaps of silicium with cellulose, is not yet certain.

81. Besides its solubility in hydrofluoric acid, one may use for the recognition of silicic acid the formation of crystals of *sodium silico-fluoride*, which are with great difficulty soluble in water. To obtain these crystals hydrofluoric acid and some sodium chloride are added to the ash and allowed to slowly evaporate. The crystals of sodium silico-fluoride which then form if silicium is present belong to the hexagonal system and represent chiefly combinations of prisms and pyramids, or of these with six-sided plates also. In stronger solutions six-rayed stars and rosettes are also observed as skeleton forms (Haushofer, I, 98).

9. Potassium, K.

82. Since ammonium cannot occur in the ash, *platinum chloride* may well serve for the recognition of potassium. The potassium-platinum chloride thus formed crystallizes in regular octahedra and cubes. According to Schimper (II, 213) the ash is dissolved in a drop of acidified water, is warmed until dry, and the reagent is added before or after cooling. But the reagent used must first be tested with much care to show that it is really free from potassium. This may be done by letting a drop of the reagent slowly evaporate on the slide.

10. Sodium, Na.

83. The *uranyl-magnesium acetate* recommended by Streng (I) serves excellently for the recognition of very small quantities of sodium. It forms with sodium a double salt of the composition $CH_3CO_2Na + (CH_3CO_2)_2UO_2 + (CH_3COO)_2Mg + (CH_3COO)_2UO + 9H_2O$. This compound, very poor in sodium and therefore formed in the presence of very small quantities of sodium, forms small colorless or very pale yellowish rhombohedral crystals, which are little soluble in water and almost insoluble in alcohol.

Since the solution of the uranium salt extracts sodium from glass vessels on long standing, Streng (III) recommends the direct addition of the solid magnesium-uranyl salt.

Schimper (II, 215) used *uranyl acetate* for the recognition of sodium, as it causes the formation, on evaporating, of sharply developed tetrahedra of sodium-uranyl acetate $(CH_3COONa + (CH_3COO)_3UO)$, of which the larger ones appear faintly yellowish. In the presence of very small quantities of sodium simultaneously with magnesium there is formed, of course, the above mentioned uranyl-magnesium-sodium acetate.

11. Ammonium, NH₄.

84. The so-called *Nessler's reagent* may be used for the recognition of ammonium, according to Strasburger (I, 74). It is prepared in the following manner: 2 grams of potassium iodide are dissolved in 5 ccm. of water, and then mercuric iodide is added to the solution while warm, until a part remains undissolved. After the fluid is cooled it is diluted with 20 ccm. of water, allowed to stand for a time, filtered, and 20 ccm. of the filtrate are diluted with 30 ccm. of a concentrated caustic potash solution. If the fluid then becomes turbid, it must be filtered again (Nickel I, 94).

In the presence of ammonium this solution takes a yellow color, and with more ammonia a brown precipitate is formed. But various organic compounds give the same reaction (Nickel I, 94).

12. Calcium, Ca.

85. Calcium occurs very often within the living plant in crystalline form, and these crystals, which are met with sometimes in the cell-sap, sometimes within the membrane, consist most commonly of calcium oxalate; crystals of calcium carbonate, gypsum, and calcium tartrate are less often observed. Besides, calcium carbonate often incrusts cell-membranes in greater quantities; and finally, calcium phosphate has been recognized in the vegetable organism. We will describe first the methods for recognizing the various calcium salts, and then the methods of recognizing the presence of calcium in the ash and in the cell-sap.

a. Calcium Oxalate, $Ca(COO)_2$.

86. Nearly all crystals which occur within the plant-cell consist of calcium oxalate. They are found partly in the cell contents, and are partly within or upon the wall. They belong partly to the tetragonal, partly to the monoclinic crystal system. Their most important forms are illustrated in Fig. 21. Here Fig. I shows a tetragonal pyramid, Figs. II and III combinations of pyramid and prism, Fig. IV a monosymmetric rhombohedron, Fig. V a rhombic plate, Fig. VI probably a combination of positive and negative hemipyramids with the basal plane, Fig. VII a combination of the rhombic plate (Fig. V) with the clinopinacoid, Fig.

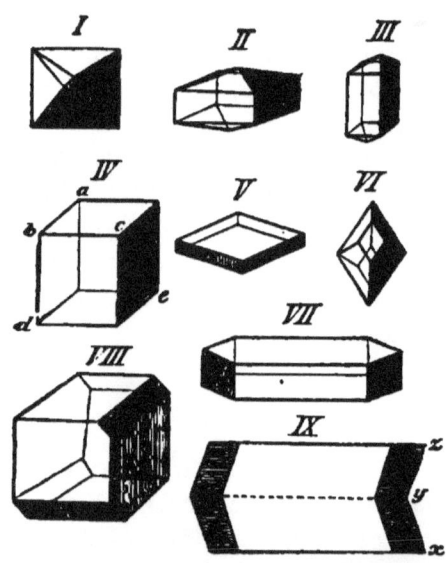

FIG. 21.—Crystals of calcium oxalate. I-III, from the spongy parenchyma of *Tradescantia discolor;* IV, from *Cycas circinalis;* V, *Musa paradisiaca;* VI, *Citrus vulgaris;* VII and IX, *Guaiacum officinale;* VIII, *Citrus medica.* IV, V, VII, IX after Holzner; VI after Pfitzner.

VIII a combination of the rhombohedron (Fig. IV) with a hemipyramid, Fig. IX a twin crystal whose angle xyz measures 141° 3′, according to Holzner (I, 34). Calcium oxalate is also especially common in the form of fine needles ("raphides") or tiny slivers ("crystal sand") on which no crystallographically determinable faces or angles can be recognized; and sphærocrystals have been seen (cf. Kohl's compilation, II, 15).

Calcium oxalate is insoluble in water and acetic acid; but in *hydrochloric acid* it is soluble, though the solution of the larger crystals, especially if they are imbedded in mucilage, does not occur at once. It is best to place the preparations in concentrated hydrochloric acid and to follow the solution with a polarizing microscope. The action of the acid can be very much hastened by warming.

With *nitric acid* calcium oxalate behaves essentially as with hydrochloric acid. It is readily soluble in the former, especially on warming.

87. By *sulphuric acid* calcium oxalate is changed into calcium sulphate (gypsum), which is little soluble in water or sulphuric acid, and separates chiefly in the form of needles. An immediate transformation of the calcium oxalate into gypsum occurs if the sections containing it are placed directly in concentrated sulphuric acid or in a mixture of equal parts of water and concentrated sulphuric acid, and heated nearly or quite to boiling. The gypsum is then formed within the same cells which formerly contained the calcium oxalate crystals; and each more or less opaque mass of sometimes plainly needle-shaped, sometimes more granular, particles of gypsum usually possesses exactly the same form as the original crystal. These crystalline conglomerates glisten brightly under the polarizing microscope.

For distinguishing calcium oxalate from calcium sulphate, Kohl has recently (II, 194) proposed a solution of *barium chloride*, which leaves the oxalate unchanged, while gypsum crystals become covered by a finely granular layer of barium sulphate. In a mixture of barium chloride and hydrochloric acid, gypsum is rapidly converted into barium sulphate, while calcium oxalate crystals disappear in the same mixture without forming any precipitate.

On treatment with *caustic potash* solution, calcium oxalate at first remains unchanged; but, as Sanio (I, 254) first observed, its crystals are suddenly dissolved after some time, usually after several hours, and new crystals are formed in the fluid, which have the form of six-sided plates whose chemical composition is not yet determined.

88. On *burning* calcium oxalate crystals, which can best be done on a cover-glass laid on platinum foil, the oxalate is changed first into calcium carbonate and then into cal-

cium oxide. The crystals preserve their original form, but become opaque and therefore appear black by transmitted light, but pure white by reflected light (or dark-field illumination). If the crystals dissolve, after the burning, in dilute acetic acid or concentrated hydrochloric acid, without the formation of gas-bubbles, this shows that an oxalate has been changed to the oxide; while the carbonate dissolves in hydrochloric acid with the liberation of carbonic acid.

89. The finding of calcium oxalate crystals can be made much easier by examination by polarized light. They are distinguished in general by their strong double refraction, which is, however, much greater in those of the monoclinic system than in those of the tetragonal system. The latter, naturally, cannot glisten in the polarizing microscope with crossed nicols, when their optical axes stand vertical.

To make visible the crystals of calcium oxalate within large organs, for example whole leaves, without cutting them into sections, these may be made quite transparent. For this purpose *chloral hydrate*, which does not attack calcium oxalate, has been used; and *phenol* can also be employed. If the pieces are heated to boiling in one of these fluids, they usually become wholly cleared in a short time. The alcoholic solution of sulphurous acid used by Wehmer (I, 218) for decolorizing leaves will certainly be of much service in many cases.

For the *preservation* of such preparations Canada balsam is best adapted. They may be transferred directly from phenol to xylol and xylol-Canada balsam. The study of these cleared preparations is best conducted by polarized light.

b. Calcium Carbonate, $CaCO_3$.

90. Calcium carbonate rarely occurs in the interior of cells, but is usually deposited in or upon the cell-wall (cf. Zimmermann I, 104).

For the recognition of the carbonic acid in calcium carbonate, *acetic* or *hydrochloric acid* may be used. After the addition of one of these, the carbonic acid is set free in bubbles, as can be directly observed under the microscope. It has been pointed out by Melnikoff (I, 30) that a concentrated acid* should be used for the recognition of small quantities of carbonic acid, and that care should be taken that it reaches the bodies to be tested as quickly as possible. Evidently, the more slowly the evolution of carbonic acid occurs, the more readily will it be absorbed by the surrounding water and carried away by diffusion without being given off in bubbles.

91. For the recognition of calcium a solution of *ammonium oxalate* acidified with acetic acid may be used.

The manner in which this solution reacts with calcium salts is largely dependent upon its strength. For example, I obtained, in sections of the leaf of *Ficus elastica*, abundant masses of crystals grown together in gland-like masses within and near the cystolith cells, by placing them in a solution containing .5% of ammonium oxalate and 1% of acetic acid. This reaction took place at once on placing the sections in the solution, which had previously been heated to boiling. The crystals thus formed are strongly doubly refractive.

But when a solution containing 10% of ammonium oxalate and 1% of acetic acid was used, the oxalate was precipitated directly in the cystoliths, which appeared quite unchanged on microscopic examination. It was only when the sections were placed in the boiling solution that the cystoliths showed a more or less granular structure on their surfaces. The presence of a crust of calcium oxalate on the cystoliths can be easily shown by placing the sections in pure 10% acetic acid, after washing out the ammonium oxalate. The still unchanged calcium carbonate incrusting the nucleus of the cystolith is then dissolved with the formation of abundant bubbles of gas, while the crust of calcium oxalate remains undissolved. By the subsequent addition of hydrochloric acid the latter is also dissolved, so that the pure cellulose skeleton of the cystolith alone remains.

---

* Concentrated HCl is best. Concentrated acetic acid cannot be used, since it dissolves calcium carbonate more slowly than dilute acid.

An aqueous 1% solution of *oxalic acid* reacts in the same way as a concentrated solution of ammonium oxalate.

92. Calcium carbonate, as well as calcium oxalate, is changed by *sulphuric acid* into gypsum. It is best in this case to use a pretty strongly dilute acid. For instance, if sections of the leaf of *Ficus elastica* are placed in 1% acid, large masses of gypsum needles are formed in the immediate vicinity of the cystoliths, formerly incrusted with calcium carbonate.

Calcium carbonate is not changed at first by *burning*, but is finally transformed into calcium oxide.

c. Calcium Sulphate, $CaSO_4$.

93. Calcium sulphate has been recognized in many Desmids by A. Fischer (I), and occurs in them chiefly in the form of tiny prisms and plates, which are sometimes enclosed in sharply-defined vacuoles, as, for instance, in the ends of the cells of *Closterium* sp. (cf. Fig. 22, *a*), or are distributed throughout those parts of the cell which contain cell-sap (cf. Fig. 22, *b*). For the microscopical recognition of gypsum, A. Fischer uses the following reactions:

Concentrated sulphuric acid leaves gypsum unchanged and undissolved when cold; *barium chloride* transforms it into barium sulphate, which is insoluble in hydrochloric and nitric

FIG. 22.—*a*, the end of a cell of *Closterium lunula*, with gypsum crystals in the apical vacuoles; *b*, median lobe of *Micrasterias rotata*. The gypsum crystals are all colored black (× 675). After A. Fischer.

acid; *burning* leaves the gypsum crystals unchanged. They are also insoluble in acetic acid, but dissolve slowly in cold caustic-potash solution, hydrochloric or nitric acid, or at once on heating.

94. Hansen (I, 10) has observed in the leaves of various

Marattiaceæ hexagonal plates, which, according to the reactions carried out by him, must consist of gypsum with an admixture of magnesium sulphate. The correctness of these views has, however, been disputed by Monteverde (II), according to whom the crystals in question consist merely of calcium oxalate. Gypsum occurs abundantly, however, according to Monteverde, dissolved in the cell-sap, and is deposited in the form of sphærocrystals after lying for months in alcohol. In the same way, a deposit of sphærocrystals consisting of gypsum is produced in *Hebeclinium macrophyllum* by alcohol, especially in the young wood-cells, according to Hansen (I, 118).

d. Calcium Tartrate, $CaC_2H_2.(OH)_2.(COO)_2$.

95. In the yellowed leaves and petioles of *Vitis* and *Ampelopsis*, Schimper (II, 238) found rhombic crystals of calcium tartrate, which sometimes reach considerable size, especially in the parenchyma of the bark and pith of the petiole of *Vitis Labrusca*. They represent largely a combination of the prism and dome, but the most various fusions also occur (cf. Fig. 23). These crystals are very slightly soluble in water, but very easily soluble in *caustic potash* solution, almost instantly so in a 10% solution, which does not attack calcium oxalate. Their behavior with *acetic acid* is also characteristic. Calcium tartrate crystals are easily soluble in dilute solutions containing about 2% of glacial acetic acid, while in the pure glacial acid, or even in a 50% solution of it, they are insoluble. In consequence of this, it may be observed, in sections to which concentrated acetic acid has been very gradually applied, that a recrystallization of previously dissolved crystals occurs.

FIG. 23.—Calcium tartrate crystals from the bark-parenchyma of a petiole of *Vitis Labrusca*, gathered Oct. 22.

Calcium tartrate crystals are doubly refractive, but this power seems to me much less than that of the monoclinic crystals of calcium oxalate. On burning they are converted into globular masses which dissolve in 10% acetic acid with the formation of bubbles.

e. Calcium Malate, $Ca(COO)_2.CH_2.CHOH$.

95a. Calcium malate is thrown down in large quantity by alcohol in the stipes of the fronds of *Angiopteris evecta*, according to Belzung and Poirault (I). It often forms prisms of considerable size, which belong to the rhombic system, and are with difficulty soluble in water, but readily so in acids. With *sulphuric acid* they form needles of gypsum. On *heating* on platinum foil they are first blackened, then show a striking increase in volume, and are finally converted into pure white lime. On being heated in the reducing flame they give off the characteristic odor of succinic acid. By the aid of Borodin's method it may be shown that they are completely insoluble in a saturated solution of neutral calcium malate.

f. Calcium Phosphate, $(CaO_2)_2(PO)_2$?

96. Calcium phosphate has been observed only in solution in the cell, except in case of globoids (cf. § 388) and of a single instance which requires confirmation (cf. Nobbe Hänlein and Councler I). It separates in the form of beautifully formed sphærocrystals in the interior of many parts of plants, after they have been placed in absolute alcohol; for instance, in the stems of *Euphorbia caput-medusæ* and *Stapelia picta*, as well as in the stalk of the frond of *Angiopteris evecta*. These sphærites are usually formed only after a considerable time (weeks or months).*

They have usually a yellowish or brownish color and are very slowly soluble in cold water. In hot water, too, they are only dissolved after a long time; at least, the solution of large sphærites was not complete after several minutes, when they had been heated to boiling in water on the slide.

With *ammonia* they behave as with water; they are only

---

* I have lately found globular or clustered bodies consisting at least chiefly of calcium phosphate in the living epidermal cells of a species of *Cyperus* (cf. Zimmermann VI, 311).

slowly soluble in *acetic acid*, but readily so in *nitric* and *hydrochloric acids*, of course without any evolution of gas.

In *sulphuric acid* they are quickly dissolved with the formation of gypsum needles. If sections are quickly heated on the slide in a mixture of two parts concentrated sulphuric acid to one part water, the masses of gypsum needles then formed show the same outline as the sphærocrystals previously present. They may be distinguished from the latter by being wholly opaque and therefore black by transmitted light, and white by reflected light. But if the sections are placed in dilute, e.g., 1%, sulphuric acid, the gypsum needles are gradually formed in the vicinity of the sphærites.

On *burning*, the calcium phosphate sphærites at first become black in consequence of the organic admixture to be mentioned in the next section, but on further heating they yield a pure white ash.

With nitric acid and ammonium molybdate, as well as with magnesium sulphate and ammonium chloride, they give the reactions for phosphoric acid (cf. § 77).

When examined with a polarizing microscope these sphærocrystals show the well-known dark cross, with crossed nicols. By the interposition of a gypsum plate it can be determined that the orientation of the optical axis is the same in them as in starch-grains and in the sphærocrystals of inulin.

In Canada balsam these sphærites may be preserved for as long as one wishes, and in glycerine gelatine at least for a considerable time.

97. The various sphærocrystals do not represent an even approximately chemically pure compound, but always contain a considerable quantity of organic substance, which often forms an amorphous nucleus at the centre of each, but is also often contained in the separate layers. It is to be ascribed to this circumstance that calcium phosphate sphærites take up pretty freely various coloring matters like methylene blue and borax-carmine (cf. Leitgeb III). The chemical composition of these organic substances is

still as uncertain as the molecular formula of the calcium phosphate contained in the sphærocrystals.

g. Recognition of Calcium in the Ash.

98. For this purpose Schimper (II, 211) recommends especially the *sulphuric acid* reaction, which may be conducted by dissolving the ash directly in about 2% sulphuric acid and then letting it dry slowly. There are thus produced, especially at the edge of the drop, crystals of gypsum which belong to the monoclinic system. Among these, plate-like crystals, whose obtuse angle (a, Fig. 24) measures 127° 31', according to Haushofer (I, 33), are especially characteristic. Besides, twin crystals are very numerous, whose edges form an angle of 104° or of 130° with each other (cf. Fig. 24). But the most various fusions are also found, on whose projecting ends pretty accurate determinations of the angles may be made.

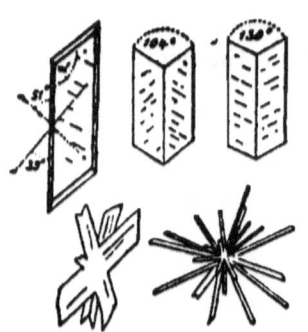

FIG. 24.—Crystals of calcium sulphate. After Haushofer.

The crystals of gypsum are also distinguished by the fact that they are transformed instantly into small needles on heating in concentrated sulphuric acid. The masses of needles preserve the form of the original crystal, but appear quite opaque if of much thickness. These needles may probably represent the anhydrite of gypsum.

If calcium is present in the ash as calcium sulphate, it will, of course, form its characteristic crystals if the aqueous solution of the ash is allowed to slowly evaporate.

h. Recognition of Calcium in the Cell-sap.

99. For the recognition of calcium in the cell-sap, Schimper employed chiefly the two following reactions:

1. On the addition of *ammonium oxalate*, calcium oxalate is formed in the cells containing calcium, in the form of

tetragonal pyramids at ordinary temperatures, but in the monoclinic form in a boiling solution.

2. Fresh sections are placed directly in a solution of *ammonium carbonate;* if calcium is present, small, strongly doubly refractive rhombohedra of calcium carbonate are formed within the cells. If the cell-sap is strongly acid, it should first be neutralized with ammonia.

13. Magnesium, Mg.

100. Schimper recommends (II, 214) the addition to the sections or to the ash, for the recognition of magnesium, of a solution of *sodium phosphate* or of *microcosmic salt* ($NaNH_4HPO_4$) reduced with a little ammonium chloride. There are then formed rhombic crystals of ammonio-magnesium phosphate ($MgNH_4PO_4$) which have in sections the form of coffin-lids, but in the ash are chiefly the X-shaped skeletons (cf. Fig. 20, § 77).

Uranyl acetate causes, if sodium is also present, the formation of the crystals of magnesium-sodium-uranyl acetate, already referred to (cf. § 83).

101. *Magnesium oxalate*, $Mg(COO)_2$. Monteverde (I) found, in the epidermis of fresh leaves of *Setaria viridis* and in dried leaves of numerous *Paniceæ*, radially striped sphærocrystals or irregular aggregates, which probably consist of magnesium oxalate. These were, according to his statements, with difficulty soluble in water, insoluble in acetic acid, and soluble in *hydrochloric, nitric,* and *sulphuric* acids, in the latter without formation of gypsum needles. After the addition of an ammoniacal solution of sodium phosphate and ammonium chloride, crystals of ammonio-magnesium phosphate were formed; after heating, these dissolved without evolution of gas; gypsum-water caused the formation of calcium oxalate crystals; and after treatment with caustic potash solution the sphærocrystals lost their striping and double refraction and became soluble in acetic acid.

Magnesium phosphate $(MgO_2)_3(PO)_2$? According to Hansen (I, 115), crystals of magnesium phosphate are precipitated in the stem of the sugar-cane by alcohol. These have partly the glandular form and are partly more or less regularly formed sphærocrystals. They are soluble in cold water with difficulty, but more easily so in hot water. They are also hardly soluble in acetic acid, but readily so in *mineral acids*, in sulphuric acid without the formation of gypsum needles. Ammonium carbonate gave no precipitate with them, but the ammoniacal solution of ammonium chloride and sodium phosphate produced a crystalline precipitate. The phosphoric acid was recognized with ammonium molybdate (cf. § 77, 1).

14. Iron, Fe.

102. Weiss and Wiesner (I) have shown microchemically that iron incrusts especially the thicker cell-membranes of the higher plants in the form of insoluble ferric and ferrous compounds, and that it also occurs in the contents of the cells. The authors mentioned used as a reagent an alcoholic solution of *potassium sulphocyanide* added directly to sections cut with a silver or platinum knife. If a red color appears at once, it shows the presence of a soluble ferric compound; but if it appears only after the addition of hydrochloric acid, the presence of a ferric compound insoluble in water is shown. In the same way sections were treated with potassium sulphocyanide and chlorine-water or nitric acid to demonstrate soluble or insoluble ferrous compounds. Large quantities of iron compounds also occur as incrustations of the membrane in various *Schizomycetes* (*Cladothrix*, *Crenothrix*, etc.) and in *Closterium*. It also forms thick crusts, in the form of ferric hydroxide, on the membranes of many *Confervaceæ* (cf. Hanstein I). For the microchemical recognition of iron, a 10% solution of potassium ferrocyanide, to which a little hydrochloric acid is added, may be used in these cases. The reagent causes the immediate formation of Berlin blue in the presence of ferric oxide. The presence of ferrous oxide may be recognized in the same way by the use of potassium ferricyanide.*

---

* A method for the recognition of organically combined iron, the so-called "masked iron," has been given by Molisch (VII). But the same

For *Leptothrix ochracea* Winogradsky (II, 268) has shown by recent investigations that the iron is first deposited in soluble form in the gelatinous envelopes, most probably as a neutral ferric salt of an organic acid. This then gradually passes over into a basic salt insoluble in water, and finally into almost pure ferric hydroxide, which is transformed by long submergence in water into a modification somewhat less soluble in hydrochloric acid.

B. Organic Compounds.

I. FATTY SERIES

1. Alcohols.

Dulcite (*Melampyrite*) $(C_6H_8(OH)_6$.

103. Dulcite has been recognized by Borodin (I) by adding one or a few drops of *alcohol* to sections of the plant under investigation, covering with a cover-glass, and allowing them to dry slowly. Dulcite then crystallizes in the form of large prismatic or irregular flattened crystals which may be distinguished from saltpeter and asparagin crystals by being insoluble in a concentrated solution of dulcite and by being transformed on heating to 190° C. into frothy dark brown masses, with complete decomposition. Dulcite crystals also differ from the very similar saltpeter crystals in dissolving without color in diphenylamine-sulphuric acid (cf. § 73).

Suitable objects for study are furnished by one-year-old stems of *Evonymus japonicus*.

---

author has recently shown (VIII) that the iron observed by him came from the caustic potash used for the reaction, and that therefore the results obtained by his method are untrustworthy. [Carl Müller (I) has still more recently concluded, not only that Molisch's proposed method is untrustworthy, but that his explanation of the source of the iron he found is equally so. Müller finds that the commercial hydroxide in stick form contains no iron, and that the iron found in solutions of caustic potash comes from the glass of the vessels in which they are contained. He believes also that the "masked" iron of Molisch is accumulated by plant specimens from the glass vessels in which they are kept, and rejects Molisch's view that most of the iron in the plant is organically combined.]

2. Acids.

a. Oxalic Acid $(COOH)_2$

104. For the recognition of oxalic acid and its soluble salts Schimper recommends (II, 215):

1. The addition of a solution of *calcium nitrate*, when crystals of calcium oxalate are formed (cf. § 99, 1).

2. The addition of *uranyl acetate* causes the formation of rhombic crystals of mostly rectangular form, which, when large, are plainly yellow and strongly doubly refractive, but whose composition is still unknown.

3. *Acid potassium oxalate*, when present in considerable quantity, is often directly recognizable in dried preparations by its crystalline form and strong double refraction on comparison with a dried solution of the same salt, as well as by the aid of Borodin's method.

b. Tartaric Acid, $C_2H_2(OH)_2(COOH)_2$.

105. Streng (III) has recommended for the recognition of tartaric acid the addition of a solution of *barium chloride* and *antimonic oxide* in hydrochloric acid. This causes the formation of rhombic plates of antimonyl-barium tartrate whose obtuse angles measure 128°.

Schimper (II, 220) recommends the use of the two following reactions:

1. The addition of *potassium acetate* produces rhombic-hemihedric crystals of the hardly soluble acid potassium tartrate.

2. Neutral solutions are treated with *calcium chloride*. There are then formed rhombic crystals of calcium tartrate which represent chiefly a combination of an elongated prism with the dome. Concerning their reactions see § 95.

c. Betuloretic Acid, $C_{36}H_{56}O_4$.

106. This acid is secreted by the trichome-glands on the leaves of *Betula alba*. It is insoluble in water, but soluble in alcohol, ether, alkalies, alkaline carbonates, and concentrated sulphuric acid, in the latter with a red coloration (cf. Behrens III, 379).

3. Fats and Fatty Oils.

107. Under the names fats and fatty oils are included, according to their consistencies, the glycerine ethers of various organic acids of high molecular weights, especially those of palmitic acid, $C_{15}H_{31}COOH$, stearic acid, $C_{17}H_{35}COOH$, and oleic acid, $C_{17}H_{33}COOH$.

But beside these, a whole series of acids still partly but little studied have been isolated from the various oils of vegetable origin (cf. Beilstein I, 427). An exact microchemical separation of these compounds is not yet possible. Even those reactions which should show whether doubtful substances belong to the group of fats still leave much to be desired in the matter of exactness, since they nearly all occur in the presence of other substances.

108. In general, however, the fatty oils show the following reactions:

They are insoluble in cold and hot water and slightly soluble in *alcohol;* but castor-oil forms an exception in being pretty readily soluble in alcohol.

They are easily soluble in *carbon bisulphide, ether, chloroform, petroleum ether,** phenol, ethereal oils* (as, e.g., clove-oil), *acetone,* and *wood-siprit* (methyl alcohol).

According to A. Meyer (II), most fatty oils are insoluble in *glacial acetic acid*, if the quantity of acid is not too great, as, for instance, when the reaction is conducted under a cover-glass.

An aqueous solution of *chloral hydrate* acts in the same way as glacial acetic acid, according to A. Meyer (II).

109. *Alcannin*, the coloring matter contained in the roots of *Alcanna tinctoria*, colors the fats deep red. The solution used for this reaction may be prepared by dissolving the commercial alcannin in absolute alcohol, adding the same volume of water, and filtering. In this solution the sections to be tested are left for one or two, or better, six to twenty-four, hours. All oil-drops then appear deeply colored; but, on the other hand, ethereal oils and resins show the

---

*[This is the *benzinum* of the U. S. Pharmacopœia.]

same reaction. The staining with alcannin may be much hastened by warming. This is especially to be recommended when one has to deal with fats which are solid at ordinary temperatures, as in the cocoa-bean. If cross-sections of this seed are heated to the boiling point in a considerable quantity of the above solution, the crystals of cocoa-butter melt and fuse into drops, which become colored deep red at once.

110. Ranvier has used (I, 97) *cyanin* (identical with chinolin blue, bleu de quinoléine) for the recognition of fats. This coloring matter is pretty easily soluble in alcohol, but practically insoluble in water, especially in cold water. On the dilution of alcoholic solutions with water, precipitates are readily formed, and I have found it most convenient to dissolve the dye in 50% alcohol and to use this solution directly for staining. Fresh material or such as has been fixed in any aqueous fixing fluid (an aqueous solution of corrosive sublimate or of picric acid, for example) may be used. It is usually sufficient for the staining to leave the objects in the above solution about half an hour. Over-stained sections may be washed out with glycerine or concentrated caustic potash solution. The permanent preservation of these preparations in glycerine-gelatine does not appear to be possible; at least, after a few months such a preparation was completely decolorized.

I can recommend as suitable objects for study old leaves of *Agave americana*, which contain large oil-drops (cf. § 364) in the leucoplasts of their epidermal cells. These are deeply colored in preparations made in the way above described, while the nuclei and chromatophores remain unstained. The only disadvantage of the method consists in the fact that the lignified and suberized membranes are also pretty deeply stained by it.

111. *Osmic acid*, commonly used in a 1% aqueous solution, colors most fats deep brown or quite black. But this reaction, which depends on a reduction, may always be checked in a short time by means of hydrogen peroxide. According to Flemming (II), the same thing may be accomplished with oil of turpentine, xylol, ether, or creosote, but it requires

usually several hours and a gentle warming for complete decolorization.

It has been shown by the researches of Altmann (I, 106) that this reaction is by no means characteristic of all fats. It is, on the contrary, suppressed in palmitic acid, stearic acid, and their triglycerides, in the mono- and triglycerides of butyrin, in lecithin, jecorin, and soap. But a strong blackening occurs with free oleic acid and olein. These two compounds are distinguished from each other by the fact that, when it is blackened by osmic acid, oleic acid is still soluble in alcohol, while olein is not.

If tannins be present in the cells to be tested for fatty oils, they should be extracted by boiling with water before the addition of the osmic acid, since they also blacken with it. Ethereal oils can be removed by heating to 130° C. (cf. § 145).

112. The *saponification* of fats under the microscope was first carried out by Molisch (I, 10, note). For this purpose he places the sections to be studied in a drop of a mixture of equal parts of a concentrated solution of potassium hydrate and a concentrated solution of ammonia. After half an hour or an hour or an even longer time, the oil-drops, "constantly losing their strong refractive power, harden into myelin-like or botryoidal bodies or into irregular masses (soaps) often consisting wholly of small crystal-needles."

Different objects seem to behave very differently in this respect. Thus, I obtained very delicate crystal-needles (cf. Fig. 25) on placing sections from the endosperm of the coffee-bean in the above-mentioned mixture of alkalies. These formed, after a few hours, around the oil-drops, which in twenty-four hours were wholly converted into crystals. I observed in places, within the crystal-aggregates, a strongly refractive rounded body, which, as examination by polarized light shows, was a sphærocrystal, and showed the familiar dark cross, with crossed nicols.

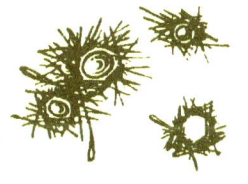

FIG. 25.—Oil-drops from the endosperm of the coffee-bean five hours after saponification with caustic potash and ammonia.

In various other objects, as, for instance, in sections from the endosperm of *Bertholletia excelsa* or from the cotyledons of *Helianthus annuus*, I obtained much larger sphærocrystals or groups of them, which were often not to be distinguished from oil-drops by ordinary illumination, but behaved quite like sphærites in polarized light.

Whether these differences are to be attributed to chemical differences between the various fats, or whether all fatty oils yield crystalline formations under the treatment described, must be determined by further researches. It is also still to be shown whether other compounds, especially many ethereal oils, do not show the same relations.

4. Wax.

113. The name wax is commonly given to the substance which covers those parts of many plants which are above ground and gives them a characteristic bright blue-green color.

Morphologically, three distinct kinds of wax coverings may be distinguished. In the first, the wax forms a complete coherent crust over the epidermis; in the second, it occurs in the form of rounded granules; in the third, in the form of small rods.

Concerning their chemical relations it may be remarked that, according to Weisner's investigations (I and II), these wax coverings contain true fats, free fatty acids, and a number of other substances. But in general the study of their chemical constitution has to do with little known compounds.

114. The wax coverings are characterized microchemically, as DeBary (I, 132) first showed, by being always insoluble in water, though they melt together into drops in boiling water, since their melting points are all below $100°$ C. They are also insoluble or hardly soluble in cold alcohol, but are always completely dissolved by boiling alcohol. In ether some of them are readily soluble; others are not soluble or very slightly so. On heating in a solution of alcannin in 50% alcohol, they run together into red drops.

Since wax is not wetted by water, the study of the various rods, granules, etc., is better conducted in cold alcohol which wets the wax without dissolving it at once, to say the least.

115. The waxy incrustations of suberized membranes, observed especially on various epidermal cells, as, for example, those of *Aloë verrucosa*, become at once visible, according to DeBary (I), if the sections are warmed under a cover-glass to near the boiling point of water. They then separate from the membrane in the form of drops. These drops are soluble in boiling alcohol and behave chemically like the wax coverings described. Wax may also be extracted from the incrusted membranes by boiling alcohol. The membranes always suffer in consequence a corresponding reduction of volume, which cannot be made good by subsequent immersion in water.

5. Carbohydrates.

116. The carbohydrates are characterized, as is well known, by the fact that they contain, besides carbon, hydrogen and oxygen in the same proportion as in water, so that their general formula may be written $C_xH_{2y}O_y$.

Of course not all organic compounds which show this empirical formula are included in the carbohydrates, and already various substances which were formerly included here have been transferred to other places in the natural system of organic compounds, after their constitution has become more exactly known. And the recent investigations of Emil Fischer (I) have introduced a more rational classification of the carbohydrates.

But these investigations have at present no significance for microchemical methods, since the certain microscopical separation of the compounds of this group is as yet possible only in very rare cases. I will therefore restrict myself here to the description of microchemical methods for recognizing some soluble carbohydrates. The solid carbohydrates, cellulose and its derivatives, as well as starch and the related compounds, will be discussed in Part III of this book (cf. §§ 242-297 and §§ 400-415).

117. But before we enter upon the special reactions of the soluble carbohydrates, two reactions common to many carbohydrates may be described. These were introduced into microchemistry by Molisch (V), and at first especially for the recognition of species of sugar. The reagents used are α-naphtol and thymol.

Molisch uses *α-naphtol* by treating sections not too thin, on the slide, with a drop of a 15-20% alcoholic solution of the compound and then adding two or three drops of concentrated sulphuric acid, so that the sections are wholly covered. In the presence of cane-sugar, milk-sugar, glucose, lævulose, maltose, or inulin, the section becomes colored a beautiful violet in a short time (about two minutes), while this reaction does not occur with inosite, mannite, melampyrite, and quercite.

If *thymol* be used in the same way intead of α-naphtol, a carmine-red color is produced.

Concerning this reaction it should be said that the concentrated sulphuric acid contained in the reagent may split off sugars from glucosides, starch, cellulose, and various other substances, and these may then give the reaction indirectly. But in the absence of soluble carbohydrates the reaction occurs much later, often only after a quarter to a half of an hour. Then, too, according to Molisch, one may reach a definite conclusion as to the presence of soluble carbohydrates by treating in the same way a fresh section and one extracted with boiling water. If the reaction occurs markedly sooner in the former, it is proved that it depends upon the presence of soluble substances.

But it has been shown by Nickel (I, 31) that, besides the compounds mentioned, a number of bodies of wholly different constitution give the same reaction, especially proteids, kreatin, and vanillin. According to Nickel, it is very probable that these reactions depend upon the fact that the sulphuric acid splits off furfurol from the compounds named.

a. Glucose, $C_6H_{12}O_6$.

118. In botanical literature the name glucose commonly includes all those kinds of sugar which precipitate cuprous oxide from an alkaline solution of copper. In most cases we have undoubtedly to do with the compound known to chemists as glucose (grape-sugar or dextrose); but there are many other substances, as, for instance, lævulose, lactose, and many glucosides, which give the same reaction; and great care should be exercised as to the significance of the copper-reaction, where such other compounds are not excluded.

The reaction named is best conducted, according to A. Meyer (IV), by first placing sections, from two to four cells in thickness, of the object to be studied, in a concentrated aqueous solution of *cupric sulphate* for a short time, then washing them in distilled water and finally placing them in a boiling solution of 10 grams of *Rochelle salt* and 10 grams of *potassium hydrate* in 10 grams of water. There are then precipitated in the cells containing glucose, vermilion granules of cuprous oxide, whose color may best be seen by dark ground illumination, as they often appear almost wholly black by transmitted light, especially with a narrow cone of rays. Cuprous oxide remains at first unchanged in glycerine, even on boiling; but, according to A. Fischer (V, 74), it is dissolved after some weeks by glycerine, as well as by Canada balsam.

119. The reaction may also be conducted on the slide in the manner recommended by Schimper (cf. Strasburger I, 73), by warming the sections, which should not be too thin, under a cover-glass in a drop of *Fehling's solution* * until little bubbles begin to be formed. Stronger heating usually causes marked changes in the cell contents.

---

* Fehling's solution may be prepared, according to Dragendorff (I, 70), by making three different solutions containing respectively, in a liter of water, 35 grams of cupric sulphate, 173 grams of Rochelle salt (sodium-potassium tartrate, $NaK(COO)_2C_2H_2(OH)_2 + 4H_2O$), and 120 grams of caustic soda. Just before use, a mixture of one volume of each of these solutions is added to two volumes of water. This mixture becomes changed in time, while the separate solutions may be kept indefinitely.

120. To recognize glucose in vessels, A. Fischer (V, 74) placed suitable pieces from branches split through the middle in a concentrated aqueous solution of cupric sulphate for five minutes, then rinsed them in water and placed them in a boiling solution of Rochelle salt and caustic soda, in which he let them boil for from two to five minutes. The cuprous oxide is then precipitated in the cells which formerly contained sugar, and the wood may be readily cut. Dried wood and alcoholic material in large pieces, whose old surfaces have been previously removed, may serve for the reaction.

b. Cane-sugar, Saccharose, $C_{12}H_{22}O_{11}$.

121. Cane-sugar is widely distributed among plants, and suitable material for study is afforded by pieces of a sugar-beet. Even on gentle boiling it cannot precipitate cuprous oxide from Fehling's solution; but on longer boiling in this solution, the cane-sugar becomes converted, in consequence of the strongly alkaline reaction of the solution, into the so-called invert-sugar, a mixture of glucose and lævulose, which reduces Fehling's solution.

For the microchemical recognition of cane-sugar, sections not too thick are placed, according to Sachs (I, 187), for a short time in a concentrated aqueous solution of *cupric sulphate*, rapidly rinsed in water, and then transferred to a solution of one part *potassium hydrate* in one part water, heated to boiling. If cane-sugar is present, there appears in the cells containing it a sky-blue color which gradually diffuses into the potash. A careful microscopic control of this reaction is recommended, since, as Sachs states, the young cell-membranes often become colored deep blue under the same treatment.

Fehling's solution may also be used for the recognition of cane-sugar, which gives a blue solution with it also. For the method of using it, see § 119.

c. Inulin, $C_{12}H_{20}O_{10}$.

122. Inulin is pretty readily soluble in water and occurs dissolved in the cell-sap of many plants. But, since it is insoluble in alcohol and glycerine, it is precipitated in the form of well developed sphærocrystals which often fill large cell-masses, when parts of plants containing inulin are preserved for a time in one of these fluids. Such sphærocrystals may be observed on examining larger parts of plants, as, for instance, halved *Dahlia* roots, which have lain several weeks, or longer, in alcohol or concentrated glycerine. After being kept for years in alcohol these are with difficulty soluble in cold water, as Leitgeb (I, 136) has shown ; while those in alcoholic material a few weeks old dissolve pretty readily in it. But in hot water even old inulin sphærites are easily soluble, so that they are readily distinguished from the otherwise similar sphærites of calcium phosphate (cf. § 96). If sections from alcoholic material of *Dahlia* roots, which always contain calcium phosphate sphærites with those of inulin, especially near the cut surfaces of the roots, be heated on the slide in water to the boiling point, the inulin sphærites dissolve almost instantly, provided the sections are not too thick, while the sphærocrystals of calcium phosphate remain for a time quite unchanged.

The inulin sphærites are also soluble easily and without residue in concentrated *sulphuric acid*, while those of calcium phosphate are transformed by it into gypsum (cf. § 96).

Fehling's solution is not directly reduced by inulin.

123. For the rapid recognition of inulin, the reagents for sugar recommended by Molisch (V, 918) may be used (cf. § 117). Thus the sphærocrystals of inulin dissolve with a very deep violet color, if the sections containing them are treated with a 10% alcoholic solution of *a-naphtol*, then with a few drops of concentrated sulphuric acid, and are then slightly warmed under a cover-glass.

But if *thymol* is added in the same way, a red color appears, according to Molisch.

Green has recommended (I) *orcin* as a reagent for inulin. The sections to be studied are saturated with an alcoholic solution of orcin and boiled in hydrochloric acid. A deep orange-red color then appears if inulin is present. Any sphærocrystals of inulin are, naturally, dissolved, and the spaces occupied by them appear red.

If *phloroglucin* is used instead of orcin, the color produced is browner.

d. Glycogen, $C_6H_{10}O_5$.

124. Glycogen, which, according to Errera's investigations, is very widely distributed in the cells of fungi, is characterized, as he has shown, by forming a colorless and strongly refractive substance within the living cells, and by becoming colored a deep red-brown with a solution of *iodine and potassium iodide*. This color disappears on warming to 50° or 60° C., and reappears on cooling. The glycogen dissolves in water if the preparation is crushed (Errera I).

Since the intensity of the color produced by iodine depends on the amount of glycogen present, by the use of iodine solutions of a given strength one may obtain some quantitative estimate of the glycogen from the color. For this purpose Errera (II) places the objects to be studied directly in a solution containing 45 grams of water, .3 gram of potassium iodide, and .1 gram of iodine. If the glycogen is present in extremely small quantity, the color will be orange rather than brown. Then a somewhat more concentrated solution (1 : 100) may be used; but it must be used very carefully.

e. Dextrine, $C_{12}H_{20}O_{10}$.

125. Dextrine is the name given to the transition product between starch and maltose. It is distinguished from the latter by being insoluble in 84% alcohol. Therefore Sachs (I, 187) proposed, for the microchemical recognition of dextrine, that sections of plants which showed the presence of copper-reducing substances with Fehling's solution should be placed in 95% alcohol for 10 to 24 hours to completely dissolve out the glucose. If the copper-reaction then still took place, he deduced the presence of dextrine. But, according to more recent investigations, pure dextrine cannot reduce Fehling's solution, and it is also doubtful whether this substance occurs in recognizable quantity within the plant (cf. Beilstein, I, 883).

6. Sulphur Compounds.

126. Of sulphur compounds of the fat group only oil of garlic and oil of mustard have been studied with reference to their microchemical recognition.

a. Garlic Oil, Allyl Sulphide, $(C_3H_5)_2S$.

Garlic oil, which occurs in almost all parts of the various species of *Allium*, gives the following reactions:

With *platinum chloride*, a characteristic yellow precipitate;
With *mercury salts*, a white precipitate;
With *palladous nitrate*, a kermes-brown precipitate;
With a 1-2% solution of *silver nitrate*, a finely granular precipitate of silver sulphide;
With *concentrated sulphuric acid*, a beautiful red color;
With *gold chloride*, a yellow precipitate.

For microchemical purposes the palladous nitrate and silver nitrate serve best, according to Voigt (I), and it is convenient to place large portions of the plants in the solution and to hasten its penetration by the aid of the air-pump. After hardening in alcohol, sections may be cut.

b. Mustard-oils, Alkylthiocarbimides, SC : N-R.

127. The mustard-oils include a group of homologous compounds which contain the atom group SC : N and an alkyl radical. The best known of them is allylic mustard-oil (allyl sulphocyanate, C_3H_5CNS), which is separated by the ferment *myrosin* from potassium myronate, which belongs to the glucosides and occurs especially in the seeds of black mustard. The reactions proposed by Solla (II) for the microchemical recognition of allyl mustard-oil have proved useless on being tested by Bachmann (VI) and Molisch (I, 33), so that we have at present no special reaction for those bodies which is microchemically applicable.

7. Amido-compounds.

128. The amido-compounds (amido-acids and acid amides) are characterized by the fact that they contain the univalent radical NH_2. They are therefore nitrogenous, and very probably play a most important rôle in the formation and transference of the albuminous materials. We have, however, trustworthy microchemical methods for recognizing only two of the amido-compounds of the fat group, *leucin* and *asparagin;* while of the aromatic amido-compounds *tyrosin* may be mentioned here (cf. § 134).

a. Leucin, Amido-caproic Acid, $C_5H_{10}NH_2.COOH$.

129. Leucin was microchemically recognized by Borodin (IV) in the leaves of etiolated specimens of *Paspalum elegans* and *Dahlia variabilis*. He made use for this purpose of its property of subliming without decomposition when carefully heated to 170° C. Dried sections were covered on the slide with a clean cover-glass and carefully heated. When leucin was present, tiny, crystalline, doubly refractive scales of this compound were deposited on the under side of the cover-glass, under this treatment. These could then be tested with a saturated aqueous solution of leucin (cf. § 71, note).

Leucin is also deposited in crystalline form, as are asparagin and tyrosin, if sections containing it are treated with alcohol and allowed to dry slowly under a cover-glass.*

b. Asparagin, Amide of Aspartic (Amidosuccinic) Acid,
$C_2H_2NH_2.CONH_2.COOH$.

130. Asparagin is soluble in water and occurs only in solution, in vegetable cells. For the recognition of asparagin according to the method first used by Borodin (II), the sections are treated with absolute alcohol under a cover-

---

* Leucin has recently been recognized by Belzung (II) in the seedlings of *Lupinus albus*. It is here precipitated in the tissues preserved in glycerine in the form of heart-shaped lamellæ, which are often aggregated into sphærocrystals.

glass and then the preparation is allowed to dry slowly. The asparagin then separates out in crystals, among which rhombic plates with an obtuse angle of 129° 18' are especially characteristic. This angle measures in the otherwise similar crystals of potassium nitrate 99° 44'.* A positive

FIG. 26.—Crystals of asparagin.

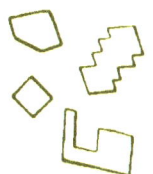

FIG. 27.—Crystals of saltpeter.

decision between asparagin and saltpeter is therefore usually possible without the actual measurement of angles, after some practice (cf. Figs. 26 and 27). The observation of the crystals is much aided by polarized light.

According to the method already employed by C. O. Müller (I), a solution of *diphenylamine* may be used to distinguish between asparagin and saltpeter, since the former is dissolved without color, while saltpeter crystals produce a deep blue color with it (cf. § 73). The behavior of asparagin crystals on heating is also characteristic. At about 100° C. they are dissolved by their water of crystallization, while at about 200° C. transformation into brown drops of froth takes place. Finally, doubtful crystals may be tested, after Borodin's method (cf. § 71, note), with a saturated aqueous solution of asparagin.

According to Leitgeb (II, 222), the crystallization of asparagin is hindered by the presence of inulin, as well as of gum, sugar, glycerine, and other viscous fluids. This author was, however, able to recognize asparagin in *Dahlia* roots, which are rich in inulin, after placing transverse slices of them a centimeter thick in 90% alcohol. After a few days he saw on the cut surfaces well formed crystals, which, after being as completely freed from inulin as possible, showed the above described asparagin reactions.

---

* But angles of 109° 56' and 118° 50' are also possible (cf. C. O. Müller I, 15); I have, however, never observed these angles, even after a very slow crystallization of pure potassium nitrate.

II. AROMATIC SERIES.

1. Phenols.

a. Eugenol, $C_6H_3.OH.OCH_3.C_3H_5$.

131. Molisch (I, 40 and 44) recommends *caustic potash* for recognizing eugenol, which occurs in the oils of pimento and clove; and he places the sections in a completely saturated solution of potassium hydroxide. There are then formed in a short time (about five minutes), from each oil-drop, numerous, often very long, columnar or needle-shaped, colorless crystals of potassium caryophyllate. Sections of cloves are especially recommended as suitable objects for their study.

The reactions with concentrated *sulphuric* or *nitric acid* are less trustworthy. The former first colors eugenol yellowish and then at once deep blue, with a tint of violet after a time, and finally brown. The latter colors it brilliant orange or brown-red.

b. Phloroglucin, $C_6H_3(OH)_3$.

132. Phloroglucin occurs in the living cell only in solution in the cell-sap. It is best recognized by means of the mixture of *vanillin* and *hydrochloric acid* proposed by Lindt (I). This is prepared by dissolving .005 gram of vanillin in .5 gram of alcohol and then adding .5 gram of water and 3 grams of concentrated hydrochloric acid. It is best to apply this reagent to previously dried sections; when, if phloroglucin be present, a light red color is produced, which later becomes somewhat violet-red. *Orcin* gives a bright blue color with a red shading; and many other related substances give no color.

Very small quantities of phloroglucin are not recognizable, on account of the red color of lignified membranes produced by the addition of hydrochloric acid (cf. §§ 255 and 257).

According to Waage (I, 253), phloroglucin in the living cell takes up methylene blue as do the tannins.

c. *Asaron*, $C_6H_2(OCH_3)_3.C_3H_5$.

133. Asaron occurs especially in the rhizome and the root of *Asarum europæum* and is here dissolved in what is very probably an ethereal oil, which almost entirely fills many parenchymatous cells of the bark. It is distilled over with water vapor and forms colorless crystals which are insoluble in water, but readily soluble in alcohol, ether, chloroform, and acetic acid.

For its microchemical recognition Borscow (I, 18) recommends especially concentrated *sulphuric acid*, which colors it first yellow, and later orange. The oil-drops containing asaron, which are found in fresh sections, give the same reaction.

2. Acids.

a. Tyrosin, p-Oxyphenyl-α-amidopropionic Acid,
$C_6H_4OH.CH_2.CHNH_2.COOH$.

134. Tyrosin has been microchemically recognized in various parts of plants, especially by Borodin (II, 816 and III, 591). He proceeded by treating sections of the objects to be tested with alcohol and then allowing them to dry slowly. Tyrosin then crystallizes out in dendritic or tufted groups of strongly refractive needles.

These are rather slightly soluble in water (in 2454 parts at 20° C., in 154 parts at 100° C., according to Beilstein, II, 1006); and they are of course insoluble in a concentrated aqueous solution of tyrosin; but Borodin's method (§ 71, note) is difficult to use with precision in this case, on account of the slight solubility of tyrosin. It is, however, characteristic of tyrosin crystals that they are colored deep red by *Millon's reagent*. They also leave a yellow residue when they are carefully evaporated with *nitric acid*. From this arises a deep red-yellow fluid, on the addition of caustic soda; and, after it dries, a red-brown crystalline deposit remains (Leitgeb II, 229).

135. According to Leitgeb (II) the crystallization of tyrosin is more or less completely prevented by the presence of inulin, in proportion to its abundance. But this author succeeded in obtaining from *Dahlia* roots large crystal-aggregates of tyrosin by placing a transversely halved root in a cylindrical vessel with the cut surface upward, and then filling the vessel with alcohol until about a third of the root projected above the fluid. The tyrosin then usually crystallized on the cut surface in a few days.

b. Ellagic Acid, $C_{14}H_8O_9 + 2H_2O$.

136. G. Gibelli (I) found in the stem and root of chestnut-trees attacked by the "malattia dell' inchiostro" [ink-disease], sphærocrystals of ellagic acid. These were soluble in water and alkalies, and dissolved in *potassium carbonate* with a yellow color, in concentrated *nitric acid* with a garnet-red color. Ferric chloride produced a greenish-black color, and silver nitrate a red-brown one.

3. Aldehydes.

Vanillin, $C_6H_3.OH.OCH_3.CHO$.

137. Vanillin occurs often on the surface of dried *Vanilla* pods in the crystalline form; but in their interior it is found only dissolved or in the amorphous condition. It is readily soluble in ether and alcohol, hardly soluble in cold water, and more readily so in hot water. It is colored blue by ferric chloride, but this reaction cannot be used microchemically, according to Molisch (I, 47). Besides, vanillin gives characteristic color-reactions with numerous organic compounds. It gives a brick-red color with *phloroglucin* or *resorcin* and sulphuric acid, a red-violet with *phloroglucin* and hydrochloric acid, yellow with *aniline sulphate*, red with *orcin* and hydrochloric acid, yellow with *metadiamidobenzol*, carmine-red with *thymol*, hydrochloric acid and potassium chlorate (cf. § 254).

Of these reagents the best adapted to microchemical use are, according to Molisch, orcin and phloroglucin. The first may be used conveniently in a 4% solution, the sections to be tested being wet with it on the slide and then treated with a large drop of concentrated sulphuric acid. If vanillin be present, the section at once becomes colored deep carmine-red throughout its whole extent. If phloroglucin is used in the same way, a brick-red color is instantly obtained with sulphuric acid; but the color produced by orcin is even more striking.

4. Quinones.

138. The quinones are characterized by the fact that the two para-atoms of hydrogen in the benzol molecule are replaced by two atoms of oxygen which are either united together by their second valence, or to the carbon atoms concerned by both valences. They are mostly colored deep yellow. It is not very probable, from the investigations already made, that they play a very important rôle in the chemistry of the plant. Nucin, emodin, and chrysophanic acid have been recognized microchemically.

a. Juglon, Nucin, Oxynaphtoquinone, $C_{10}H_6O_2.OH$.

139. Nucin has been recognized in the cell-sap of the parenchyma of the outer husk (pericarp) of the fruit of *Juglans regia* by O. Herrmann (I, 183). He used for this purpose a solution of ammonia, or better, the fumes of ammonia, which at first color the nucin a brilliant purple; but this color gradually passes into brown.

b. Emodin, Trioxymethylanthraquinone, $C_{11}H_4O_2.CH_3.(OH)_3$.

140. Bachmann (I) observed in the lichen *Nephroma lusitanica* that small yellow crystalline granules adhere to the exterior of the hyphæ of the pith, which agree in their microchemical reactions with the emodin previously produced macrochemically only from the rhubarb root and the fruits of *Rhamnus frangula*. They are dissolved, like chrysophanic acid, by *caustic potash* and soda solutions with a red color. *Lime* and *baryta waters* color them dark red but do not dissolve them. But they are distinguished from chrysophanic acid by being readily soluble in *alcohol, glacial* *acetic acid*, and *amyl alcohol*, and in dissolving in concentrated *sulphuric acid* with a saffron-yellow color; while chrysophanic acid is very slightly soluble in the first three reagents and dissolves in sulphuric acid with a rose-red color. According to Fr. Schwarz (II, 251), emodin is also distinguished from chrysophanic acid by dissolving in *ammonium carbonate* with a red color, while the latter remains unchanged in this reagent.

c. Chrysophanic Acid, $C_{14}H_4O_2.CH_3.(OH)_2$.

141. Chrysophanic acid has been observed especially in the lichen *Physcia parietina* and in the roots of various *Polygonaceæ*. In the latter it occurs, according to Borscow (I), in the form of yellow granules imbedded in the cytoplasm. But in *Physcia* it is in the form of crystalline granules adhering to the membrane, as Fr. Schwarz (II, 262) has shown, in opposition to Borscow (I). These are very small and can be plainly recognized only by strong magnification. In polarized light they glisten brightly with crossed nicols. They are characterized by the deep crimson color which they take with *caustic potash* or *ammonia;* while *lime* and *baryta water*s color them dark red, without dissolving them (see also § 140).

5. Hydrocarbons, $(C_{10}H_{16})x$.

142. A large group of various vegetable substances is included by Beilstein (III, 279) under this title. They represent either terpenes of the composition $C_{10}H_{16}$ or polymerization products of these, or at least are very nearly related to them. Beside the true terpenes there are included here especially the ethereal oils, caoutchouc and gutta percha, the resins and balsams. The chemical constitution of some of these compounds has been very little studied, and doubtless many of the substances placed here will in time be transferred to other parts of the natural system, especially many of the so-called ethereal oils. Thus, the chief constituent of the so-called ethereal oil of the species of *Allium*,

garlic-oil, is a relatively simple compound, allyl sulphide (cf. § 126).

Since we have no microchemically applicable reactions for most of these compounds, one must be content in most cases to recognize them as members of this group, which may conveniently be called the group of the terpene-like compounds, except where macroscopic studies of the substances contained in the plant concerned afford points of vantage which may be turned to account microchemically. In many cases even this recognition cannot be made with certainty, for lack of a completely trustworthy reaction for the group.

143. If we omit caoutchouc and gutta percha, most of these compounds are characterized by being *strongly refractive* and quite or almost insoluble in water. They are, however, soluble in the solvents of fats, like ether, chloroform, carbon bisulphide, benzol, and ethereal oils. They are also mostly readily soluble in cold *alcohol*. Like the fats, they are deeply colored by *alcannin*, and the process is just the same as for the fatty oils (§ 109).

Besides these, the following special reactions may be described :

a. Ethereal Oils.

144. The ethereal oils are characterized microchemically by the fact that they may be obtained from the plants by distillation with water vapor, and that they leave on paper a greasy spot which disappears after a time ; and for their microchemical recognition their *volatility* may be utilized. To test this, sections of the parts concerned may be boiled in water for a time and examined with the microscope before and after the boiling.

According to A. Mayer (II), it is sufficient for the removal of all ethereal oils to heat the uncovered sections for ten minutes in a drying oven up to 130° C., since the fatty oils remain unchanged under this treatment. Otherwise the ethereal oils agree with the fatty oils in being browned or blackened by *osmic acid* and deeply stained by *alcannin* or *cyanin* (cf. §§ 109-111).

Most ethereal oils are also readily soluble in glacial *acetic acid* and in an aqueous solution of *chloral hydrate*.

[Mesnard (I) has used the following method for distinguishing the ethereal from the fatty oils. He cemented two glass rings of different sizes concentrically to a slide, the inner ring being also lower than the outer. The space between the two rings was filled with strong hydrochloric acid, and the sections to be examined were placed on the under surface of a cover-glass resting on the outer ring, in a hanging drop of glycerine containing a large proportion of sugar. Other sections may be placed, for longer exposure, on a small cover that rests on the inner ring. In this apparatus the HCl is taken up with water by the glycerine, and, after a time, the ethereal oils present exude in golden-yellow drops, which later disappear. This exudation in drops never occurs with fatty oils.]

It may be observed here that J. Behrens (I) has seen on the glandular hairs of *Ononis spinosa* a strongly refractive secretion which became colored deep red, even in very dilute aqueous solutions of fuchsin. Nothing definite can be said as to the chemical composition of this secretion, but it is probable that it belongs in the category of ethereal oils.

b. Resins and Terpenes.

145. The Unverdorben-Franchimont reaction with *copper acetate* may be used as a special reagent for resins and terpenes. Large pieces of the parts of plants to be examined may be placed in a concentrated aqueous solution of the salt named, and studied after not less than about six days. The resins then appear colored a beautiful emerald-green. The copper acetate may be removed before cutting by washing with running water. Pieces so treated may be preserved in 50% alcohol; and microscopic preparations retain their beautiful color in glycerine-gelatine.

Alcannin has also been much used (cf. § 109) in the study of the resins. Preparations treated with this stain do not appear, however, according to the author's experiments, to be capable of preservation in glycerine-gelatine.

In connection with the resins may be mentioned the following compounds not yet well studied macrochemically.

α. Fungus-gamboge.

146. Zopf (V, 53) denominates as fungus-gamboge a yellow resinous substance which corresponds in all its determined characters with the gamboge-yellow which is the chief constituent of gamboge. It occurs especially in the sporophore of *Polyporus hispidus*, chiefly deposited in the membranes, but also as an excretion outside of the membranes and in the cell-contents. For its microchemical recognition Zopf uses a solution of *ferric chloride*, which colors fungus-gamboge, as well as membranes containing it, olive-green or blackish brown. It is also insoluble in water, but readily soluble in alcohol and ether; it is dissolved with a red color by concentrated nitric or sulphuric acid, and is precipitated from these solutions in yellow flocks on the addition of water.

β. Retinic Acid from *Thelephora* sp.

147. Zopf has prepared (V, 77) a retinic acid from various species of *Thelephora*, which occurs partly in the cell-contents and is partly deposited in or on the walls. It is insoluble in cold or hot water, soluble in alcohol, ether, methyl alcohol, petroleum-ether, chloroform, benzol, carbon bisulphide, and turpentine. It is dissolved with a bluish-red color by concentrated *sulphuric acid*, and thrown down again with a greenish-yellow color on the addition of much water.

γ. Retinic Acid from *Trametes cinnabarina*.

148. The retinic acid prepared by Zopf (V, 88) from the above-named fungus agrees fully with the previous one in its behavior with solvents; but is distinguished from it by being dissolved by concentrated *sulphuric acid* with a yellowish or reddish brown color. It differs from fungus-gamboge in taking no olive-brown color with ferric chloride.

δ. Retinic Acid from *Lenzites sepiaria.*

149. According to Bachmann (II, 7), opaque globules or granules of a retinic acid are found on the membranes of *Lenzites sepiaria* in many places. These are quickly dissolved by a watery or alcoholic solution of *caustic potash* or soda with a dark olive-green color.

6. Glucosides.

150. The name glucosides is given to a group of substances resembling compound ethers, widely distributed in the vegetable kingdom, which are characterized by being decomposed by various reagents, especially acids, alkalies, or ferments, into a species of sugar and one or several other compounds. Usually this species of sugar is glucose. But there are commonly included in the glucosides compounds which do not yield a true sugar, like, e.g., phloretin, which forms phloroglucin instead of glucose (cf. Beilstein, III, 322.)

The glucosides which yield glucose on decomposition may be microchemically recognized by the aid of Fehling's solution (cf. § 119) under some circumstances, even when they do not directly reduce a copper solution, but only after the glucose has been separated from them, as by warming with dilute sulphuric acid. There are also many glucosides which directly reduce Fehling's solution.

In their other relations the glucosides show great differences, and no reactions common to the entire group are yet known. We must therefore confine ourselves here to the special description of some glucosides for which special reactions, microchemically applicable, have been suggested.

a. Coniferin, $C_{16}H_{22}O_8$.

151. Coniferin has been macrochemically produced from the cambial juices of various *Coniferæ*. But, although coniferin presents a considerable number of color-reactions, it has not yet been successfully recognized within the cells by microchemical means. Yet the various reactions for coniferin succeed easily with all lignified cell-membranes, and it is usually assumed that they contain coniferin (cf. §§ 254-256).

b. Datiscin, $C_{21}H_{22}O_{12}$.

152. Datiscin has been recognized microchemically by O. Hermann (I, 9), especially in the cell-sap of the bark-parenchyma and as a deposit in the cell-walls of the thick-walled cells of the wood and bark of *Datisca cannabina*. This author used principally *lime* and *baryta waters* which give a pure yellow solution with datiscin, and cause a deep yellow coloring of all cells containing it; while the addition of acetic acid or dilute hydrochloric acid instantly causes the color to disappear. Besides, datiscin gives yellow precipitates with *lead acetate* or *zinc chloride*, a greenish one with *cupric salts*, or a dark brownish-green one with *ferric chloride*.

c. Frangulin, $C_{20}H_{20}O_{10}$.

153. Frangulin forms a yellow crystalline mass, which is insoluble in water, but dissolves in *alkalies* with a cherry-red color. According to Borscow (I, 33), it occurs inside the parenchymatous elements of the stem of *Rhamnus frangula*, united with small starch-granules, which are colored blue by a solution of iodine, and blood-red by caustic potash or ammonia (?).

d. Hesperidin, $C_{21}H_{26}O_{12}$ (or $C_{50}H_{60}O_{27}$?).

154. Hesperidin is dissolved in the cell-sap within the living plant; but it is deposited, like inulin, in tissues which have lain for some time in alcohol or glycerine, in the form of sphærocrystals. According to Pfeffer (III), unripe oranges are especially favorable objects for study.

The sphærocrystals of hesperidin are distinguished from

those of inulin by having usually a much less rounded form than the latter, and by showing their composition from separate acicular crystals much more clearly. The latter is the case especially in the smaller sphærites (cf. Fig. 28, *a* and *b*); the larger ones commonly contain in the middle a more homogeneous nucleus, as Fig. 28, *c* shows in optical median section.

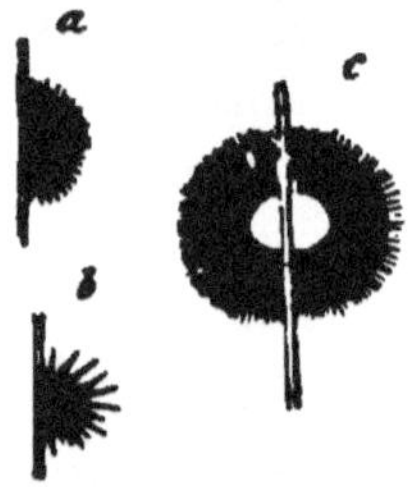

FIG. 28.—Hesperidin crystals from alcoholic material of unripe fruit of *Citrus Aurantium*.

The hesperidin crystals readily dissolve in an alcoholic or aqueous solution of *caustic potash*, forming a yellow fluid, but are not noticeably soluble in cold or hot water or in dilute acids; while inulin is at once completely dissolved by boiling water.

The sphærocrystals of hesperidin are also soluble in boiling concentrated *acetic acid, ammonia,* and *soda solution,* but are insoluble in ether, benzol, chloroform, carbon bisulphide, and acetone. They are completely destroyed on heating to a red heat.

The observation and preservation of these crystals in Canada balsam may be readily accomplished after clearing in oil of cloves. Canada balsam preparations are especially adapted to their study by polarized light, with which they give the same appearance as inulin sphærites.

e. Coffee-tannin, $C_{16}H_{18}O_9$.

155. Coffee-tannin, according to Molisch (I, 9), shows the following reactions, which may easily be followed under the microscope on sections of the endosperm of the coffee-bean. With *ferric chloride* it is colored dark green, with *ammonia* and *caustic potash* deep yellow. If sections are allowed to dry with a drop of ammonia, they finally take a green color, which at once changes to red on moistening with concentrated sulphuric acid. An abundant precipitate is formed with *lead acetate.*

f. Potassium Myronate, $KC_{10}H_{18}NS_2O_{10}$.

156. Potassium myronate, which occurs in the seeds of many *Cruciferæ*, is split up by the ferment myrosin into the allylene mustard-oil, glucose, and potassium sulphate. For its microchemical recognition Guignard (II) first treats sections with alcohol, which dissolves out any fatty oil and makes the myrosin inactive, while potassium myronate is almost insoluble in it. If the sections are then placed in an aqueous extract of white mustard-seed, which is very rich in myrosin, there is formed in the cells containing myronic acid the allylene oil of mustard, which Guignard recognizes by the aid of tincture of Alcanna (cf. § 109).

g. Phloridzin, $C_{21}H_{24}O_{10}$.

157. O. Herrmann (I, 21) uses for the microchemical recognition of phloridzin, solutions of *ferric chloride* and *ferrous sulphate*. The former produces a dark red-brown solution, the latter a yellow-brown precipitate. These reactions have led to trustworthy results only with *Pirus Malus;* in case of the pear, cherry, and plum, they are too much masked by the presence of large quantities of tannin which gives a green color with iron salts.

h. Ruberythric Acid, $C_{26}H_{28}O_{14}$.

158. Ruberythric acid forms the chief constituent of the so-called madder dye in the roots of *Rubia tinctorum*, and is decomposed, on boiling with hydrochloric acid, into glucose and alizarine (cf. Husemann I, 1387). It has a yellow color and, as Naegeli and Schwendener (I, 502) have shown, is exclusively dissolved in the cell-sap in young roots, while the membranes are still quite colorless. In older roots the membranes are, however, colored, even in such as are still living, as Naegeli and Schwendener showed by plasmolysis of the cells.

Caustic potash solution colors the mixture of coloring matters in *Rubia tinctorum* purple-red, *acids* color it orange, *ferric chloride*, orange to brown-red. When the roots dry, it is decomposed with the formation of red masses, probably through the agency of a ferment.

i. Rutin, $C_{42}H_{50}O_{26}$.

159. Rutin always occurs in the cell-contents, according to O. Herrmann (I, 30). He recommends for its microchemical recognition *ammonia* or *lime-water*, which form deep-blue solutions with rutin, becoming brown on exposure to the air.

k. Saffron-yellow, Crocin, $C_{44}H_{70}O_{28}$.

160. The crocin contained in saffron is readily soluble in water, less so in alcohol, and very slightly so in ether. It dissolves in concentrated *sulphuric acid* with a blue color which soon becomes violet and finally brown; in concentrated *nitric acid* with a deep blue color, quickly becoming brown (Beilstein III, 357). According to Molisch (I, 57), the last two reactions may well be used microchemically.

l. Salicin, $C_{13}H_{18}O_7$.

160a. Salicin occurs especially in the bark of willows and poplars. It is colored a beautiful red by concentrated *sulphuric acid* (Rosoll I, 8).

m. Saponin, $C_{19}H_{30}O_{10}$.

161. Saponin is dissolved in the cell-sap within the living plant, according to Rosoll (I, 11), but is precipitated, on drying, in the form of amorphous lumps within the cells, which may be readily observed in oil or glycerine. They are dissolved in water or very dilute alcohol, but may be precipitated from the solution by alcohol or ether.

With concentrated *sulphuric acid* saponin gives at first a yellow, then a bright red, and finally a blue-violet color. To prevent confusion with Raspail's protein reaction (cf. § 227), which also differs in the colors produced, control sections may be boiled in water to extract the saponin, and then treated in the same way with concentrated sulphuric acid.

n. Solanin, $C_{42}H_{70}NO_{16}$.

162. According to Wothtschall (I), who has tested a great number of reactions as to their applicability, only the three following are suited for the microchemical recognition of solanin.

1. *Mandelin's Reaction.*—The reagent for this test should be freshly prepared by dissolving one part of *ammonium vanadate* in 1000 parts of a mixture of 98 parts of concentrated sulphuric acid with 36 parts of water. This is added directly to the sections to be studied. In the presence of solanin the following colors appear in order: yellow, orange, purple-red, brownish-red, carmine, raspberry-red, violet, blue-violet, pale greenish blue; finally all color disappears. The time in which this scale of color is gone through is chiefly dependent on the concentration of the solutions present, but is always several hours. If any fatty oils are present in the preparation, which would also give color reactions with concentrated sulphuric acid, they may be removed by preliminary submersion of the sections in ether, in which solanin is practically insoluble.

2. *Brandt's Reaction.*—.3 of a gram of *sodium selenate* is dissolved in a mixture of 8 ccm. of water and 6 ccm. of concentrated sulphuric acid. After the addition of this reagent to the sections to be studied, the preparation is gently warmed by moving it over a small flame. As soon as the color appears the warming must be stopped. In the presence of solanin there appears first a raspberry-red color, which gradually passes into a currant-red, which soon becomes paler and more brownish yellow, and finally quite disappears.

3. *Sulphuric Acid.*—On the addition of concentrated sulphuric acid, solanin gives at first a bright yellow color, which gradually becomes redder, then takes a violet shade, gradually pales, passes into greenish, and finally quite disappears.

For the preservation of plants containing solanin, Wohtschall (I, 186) recommends simply drying them.

o. Syringin, $C_{17}H_{24}O_9 + H_2O$.

163. Syringin occurs, according to Borscow (I, 36), in branches of *Syringa vulgaris* within the thick-walled elements of the phloëm and of the xylem, as well as in the medullary rays, but only in the cell-walls. For its microchemical recognition this author uses a mixture of one part concentrated *sulphuric acid* and two parts water. This is added to delicate longitudinal or transverse sections and colors the walls containing syringin first yellowish green, after a few minutes blue, and finally violet-red.

p. Glucoside (?) *from the Stimulus-conducting Tissue of Mimosa pudica.*

164. Haberlandt (I, 17) has observed that when the stem or petiole of *Mimosa pudica* is cut, the drops of fluid which escape leave, on drying, a considerable quantity of a crystallized substance whose composition has not yet been made out, but which should be for the present included among the glucosides, on account of its reactions.

The crystals of this substance are always colored pale brownish and show very various forms. They sometimes form large prisms or cross-shaped twins, sometimes glandlike masses or dendritic bodies; and sphæritic formations have been seen. The material identity of these different formations is shown by the fact that they all dissolve with a red-violet color in *ferric chloride*. They are also soluble in water, but very slightly so in alcohol, and quite or almost quite insoluble in ether. They are dissolved by concentrated *sulphuric acid* with a yellow-green color, which becomes red-brown on warming. Dilute sulphuric and hydrochloric acids throw down from the aqueous solution a finely granular white precipitate, which is soluble in alcohol. *Ferrous sulphate* causes a deep rust-red color. *Lead-acetate* produces a heavy yellow precipitate in the aqueous solution, which is soluble in acetic acid. The substance does not directly reduce Fehling's solution, but does so after heating with dilute sulphuric acid.

Since the isolation of this questionable substance has not yet been accomplished, it must remain doubtful, in case of several of the reactions described, how far they are influenced by other substances also contained in the drops of fluid.

7. Bitter Principles and Indifferent Substances.

a. Calycin, $C_{18}H_{12}O_5$.

165. Calycin occurs in the form of small yellow crystals deposited on the membranes of *Calycium chrysocephalum* and various other lichens. For its microchemical recognition it is best, according to Bachmann, to treat with *glacial acetic acid* a bit of the lichen under examination, which has been rubbed as fine as possible, then to bring the whole together into a drop and let it evaporate. The calycin crystallizes out in long, acicular, strongly doubly refractive crystals.

It is also characteristic of calycin that it is not dissolved by caustic potash solution and suffers no change of color by it.

b. Spergulin, $(C_5H_7O_2)x$.

166. Harz (II) has isolated from the seed-coats of *Spergula vulgaris* and *S. maxima* a strongly fluorescent substance which he calls spergulin. It is readily soluble in absolute and dilute *alcohol* and appears colorless or faintly greenish or olive-brown in this solution, showing an intense, dark-blue fluorescence, while a beautiful emerald-green fluorescence appears after the addition of a small quantity of alkali. It cannot be obtained in crystalline form from this solution.

It is also characteristic of spergulin that it dissolves in concentrated *sulphuric acid* with a beautiful dark blue color. It is readily soluble in methyl alcohol, less so in amyl alcohol, and hardly so in petroleum and ether.

It is insoluble in fatty and ethereal oils, in benzine, carbon bisulphide, chloroform, phenol, cold and hot water, and dilute organic and inorganic acids.

Since the membranes of the strongly thickened outermost cell-layer of the seed-coat dissolve in concentrated sulphuric acid with a deep-blue color, Harz concludes that spergulin is contained in the membranes themselves.

8. Coloring Matters.

167. The compounds united under this head do not form a group of chemically related substances, but there are included here all colored substances whose chemical constitution is still too little known to enable them to be placed in their proper position in the natural series. Concerning most of them, then, our knowledge is extremely incomplete, and an actual isolation and quantitative analysis has been carried out with any exactness upon very few of them.

It cannot be my duty to enumerate all the coloring matters heretofore extracted from various parts of plants. It seems rather that I should restrict myself to those concerning whose chemical relations we have some trustworthy information, and which, especially, are microchemically recognizable with some certainty.

I have grouped the coloring matters according to the place in the living plant where they occur.

a. The Pigments of the Chromatophores.

168. The different colorings of the chromatophores, according to our present knowledge, if we omit for the moment the algæ which are not green, are to be referred to a relatively small number of coloring matters. But these are still too little studied to make a completely certain limitation possible. Four coloring matters may be distinguished with some certainty, and I will here confine myself to their brief description. A special discussion of the very abundant literature of chlorophyll does not seem desirable here, since it contains almost exclusively the results of macroscopic investigations and has hitherto produced very few results of physiological value.

After the discussion of these four pigments, the coloring matters of the chromatophores of the algæ which are not green will be taken up.

α. Chlorophyll-green.

169. The chloroplasts or chlorophyll-grains seem to owe their color in all cases to one and the same coloring matter, to which is commonly given the name chlorophyll-green, or chlorophyll. It is macrochemically distinguished by its strong red fluorescence and by its absorption spectrum, in which may be distinguished four bands, besides the end absorption beginning in the blue. Of these the strong band in the red is especially characteristic. Since no other green coloring matter has yet been observed in the chromatophores, it is only in case of feebly colored chloroplasts that any doubt can arise as to the presence of chlorophyll in microscopical observation. In such cases the so-called *hypochlorin* or *chlorophyllan reaction* has been used for the recognition of chlorophyll-green. According to A. Meyer (II), this reaction is best conducted by treating the sections under a cover-glass with glacial acetic acid. There are then extruded on the surfaces of the chromatophores, sometimes crystalline, sometimes amorphous masses which contain decomposition-products of chlorophyll. For further reactions of hypochlorin, see A. Meyer, II.

β. Carotin, Chlorophyll-yellow.

170. Carotin was first prepared from the roots of *Daucus Carota*, where it occurs in the form of rhombic plates or variously shaped crystalline formations (cf. § 356). The same substance also occurs in the orange or red chromatophores of many flowers and fruits. But, according to the researches of Arnaud (II), it is constantly to be met with within the chloroplasts, which, according to Immendorff (I), contain no yellow or yellowish-red coloring matter except carotin. Carotin would thus be identical with the coloring matters called chlorophyll-yellow, xanthophyll, erythrophyll, chrysophyll, etc. According to Immendorff (I, 516), carotin also occurs in the chromatophores of etiolated parts of plants and is the cause of the autumnal yellowing of leaves.

171. The elementary composition of carotin corresponds to the formula $C_{26}H_{38}$, according to Arnaud (I); it would therefore clearly constitute an hydrocarbon. According to Immendorff (I, 510), it is precipitated in shining, deep-red crystals by alcohol, from its solution in carbon bisulphide, which it colors deep blood-red. These crystals show, when undecomposed, a strong dichroism (cf. § 356, note); but on lying in the air they take up oxygen and pass over gradually into a colorless compound readily soluble in alcohol. On warming they first take a brick-red color, and above 160° C. they melt.

Courchet has prepared very variously shaped crystals of carotin from various parts of plants which showed, like those observed naturally, sometimes a red, sometimes a more yellow color. It is still unknown to what these differences in color, which are to be seen in one and the same cell in the carrot, are to be referred. It is also yet to be determined whether all the coloring matters called carotin are identical; for it seems probable that we have to do here at least with a group of nearly related coloring matters.

172. The following reactions may serve for the microchemical recognition of carotin: With a solution of *iodine* (e.g., aqueous solution of iodine and iodide of potassium), it is colored greenish or greenish blue; with concentrated *sulphuric acid*, first violet and then indigo-blue; from its deep blue solution in concentrated sulphuric acid it is precipitated in green non-crystalline flakes, on the addition of water, according to Immendorff (I, 509).

Carotin is also, according to Arnaud (I), insoluble in water, almost so in alcohol, very slightly soluble in ether, more readily so in benzine, and most so in *chloroform* and *carbon bisulphide*.

These solutions, even when from carmine-red crystals, are colored yellow or orange-yellow, according to their concentration, while the solution of carotin in carbon bisulphide, as already stated, is always blood-red.

The absorption spectrum of carotin shows, according to Immendorff (I, 510), two bands in the blue and absorption of the violet.

γ. Xanthin.

173. Xanthin occurs in the yellow chromoplasts, always in amorphous form, and especially in small granules (grana) (cf. § 357). Its alcoholic solution leaves, on evaporation, according to Courchet (I, 349), a wholly amorphous, resin-like mass. It is insoluble in water, little soluble in ether, chloroform, and benzine, but more so in *alcohol*. With concentrated *sulphuric acid*, the isolated pigment, as well as the chromoplast, takes first a greenish, then a blue color; with *iodine*, best used in the form of the solution with potassium iodide, it becomes green.

δ. Coloring Matter of *Aloë* Flowers.

174. A coloring matter whose reactions differ essentially from those of xanthin and carotin has been recognized by Courchet (I) in the chromoplasts of the flowers of *Aloë*. It is insoluble in ether and chloroform, but readily so in *alcohol*, with a currant-red color. On dilution its solution becomes rose-red; benzine takes up but little of it; and it has not yet been obtained in crystalline form. It becomes yellowish green with *sulphuric acid*, and about the same with *hydrochloric acid; nitric acid* first turns it yellow and then decomposes it. *Caustic potash* colors the granules of the coloring matter orange and makes them run together; *iodine* colors them dirty yellow.

This substance has not yet been observed in other plants.

ε. Coloring Matters of the Chromatophores of the *Florideæ*.

175. In the chromatophores of the *Floridcæ* there occur a red pigment soluble in water and a green pigment soluble in alcohol. The latter is almost certainly identical with

chlorophyll. The pigment soluble in water is at present commonly called phycoerythrin.

Phycoerythrin is, according to Schütt (II and IV), insoluble in alcohol, ether, benzole, benzine, carbon bisulphide, glacial acetic acid, and fatty oils. Its saturated aqueous solution appears by transmitted light dark bluish red, by reflected light more or less yellow, on account of its deep orange-yellow fluorescence. Its absorption-spectrum is distinguished from that of chlorophyll, according to Schütt's investigations (II), especially by the fact that its maximum of brilliancy is at the point in the red where the strongest absorption-band of chlorophyll occurs.

ζ. Coloring Matters of the *Phæophyceæ*.

176. Millardet has recognized three pigments in the chromatophores of the *Phæophyceæ*: chlorophyllin, phycoxanthin, and phycophæin. The first two of these are, according to Hansen (II), identical with chlorophyll-green and chlorophyll-yellow; that is, with chlorophyll and carotin.

Phycophæin is readily soluble in water, especially in hot water; its saturated solution has a deep red-brown color and shows with the microspectroscope a regular increase of absorption from the red to the blue end of the spectrum without any absorption-bands. It is also little soluble in dilute, and insoluble in absolute, alcohol, as also in ether, carbon bisulphide, benzol, benzine, and fatty oils. It is more or less completely precipitated by *acids* from its aqueous solution, incompletely so by *caustic soda*, and not at all by ammonia and salts of the alkalies. Salts of the alkaline earths precipitate it (Schütt III).

η. Coloring Matters of the *Cyanophyceæ*.

177. Three different pigments have been isolated by Reinke (II, 405) from an *Oscillatoria*, which he calls chlorophyll, phycoxanthin, and phycocyanin. Of these the first must certainly be identical with ordinary chlorophyll, but whether the phycoxanthin is to be identified with carotin is

not yet certain, since, according to Reinke's statements, this is not very probable.

Phycocyanin is soluble in water, and in this solution has a bright blue color with a red fluorescence. Its absorption-spectrum shows four bands, according to Reinke.

θ. Coloring Matters of the *Diatomaceæ*.

178. The chromatophores of the *Diatomaceæ*, which are colored yellowish brown in the living state, are colored greenish by *alkalies* and dilute *acids*, and by concentrated *sulphuric acid* a beautiful verdigris-green, as Naegeli has shown. They contain at least two pigments, one of which is surely identical with chlorophyll, while the other is often called phycoxanthin, but more appropriately *diatomin*.

Diatomin is characterized by being soluble in dilute alcohol with a brownish-yellow color. It is not probable, from the investigations already made, that it is identical with carotin (cf. Askenasy I, 236 and Nebelung I, 394).

ι. Coloring Matters of the *Peridineæ*.

179. The *Peridineæ*, which have recently been commonly included among the Algæ, are distinguished, as Klebs (V, 732) has shown, by the possession of true chromatophores. These contain, according to Schütt (I), three pigments, which he calls Peridine-chlorophyllin, peridinin, and phycopyrrin. Of these, Peridine-chlorophyllin is either identical with, or very closely related to, chlorophyll.

1. *Peridinin* is insoluble in water, but readily dissolves in *alcohol*, giving a fluid of the color of red wine. It is also readily soluble in ether, chloroform, benzol, carbon bisulphide, and glacial acetic acid. Its absorption-spectrum shows, according to Schütt (I), the Band I of chlorophyll between B and C, and is also marked by a strong absorption in the green-yellow.

2. *Phycopyrrin* (from πυρρός, red-brown) is soluble in water, giving a dark brown-red fluid. It is also easily soluble in alcohol, ether, bisulphide of carbon, and benzol.

The absorption-spectrum of the aqueous solution has, according to Schütt, a certain similarity to the chlorophyll spectrum, in that it shows Band I.

b. *Fatty Pigments or Lipochromes.*

180. At present, all those yellow or red pigments which are colored blue by *sulphuric* or *nitric acid*, and green by solutions of *iodine* with potassium iodide, are called fatty pigments or lipochromes. They are mostly dissolved in fatty substances within the living cell.

Zopf (IV) has recently investigated the relations of the lipochromes, especially toward sulphuric acid, and has proved that, in this reaction, deep-blue crystals often occur, which he calls *lipocyanin* crystals. These are formed especially when rather dilute sulphuric acid is added to the residue from an evaporated solution of a lipochrome upon the slide. Zopf also succeeded in obtaining lipocyanin crystals directly from various organs by letting the sections dry before treating them with sulphuric acid.

181. According to their chemical relations, the already described pigments, *carotin* and *xanthin*, belong to the lipochromes (cf. §§ 170 and 173). There also belongs here the red pigment prepared by Cohn from the cells of *Hæmatococcus*, which he calls *hæmatochrome*. The pigment called *chlororufin* by Rostafinski (I) is also to be included here. Zopf and Bachmann have prepared various lipochromes from different fungi (cf. Zopf II, 414). Finally, the *bacteriopurpurin* prepared by Lankaster from various red Bacteria is to be placed among the lipochromes, according to Bütschli (I, 9), whose studies indicate that it is identical with Cohn's hæmatochrome.

Whether all these pigments are nearly related chemically, and how far they are identical with each other, or, on the contrary, consist of groups of more or less different pigments, must be determined by further investigations.

c. Other Coloring Matters dissolved in Fats or Ethereal Oils.

182. Of the other pigments which from their chemical reactions do not belong to the lipochromes, but occur dissolved in fats or ethereal oils in the living plant, only curcumin has been microchemically investigated.

Curcumin, $C_{14}H_{14}O_4$.

Curcumin occurs, according to O. Herrmann's investigations (I, 24), in the fresh rhizome of *Curcuma amata* within the parenchymatous cells of the fundamental tissue, dissolved in an ethereal oil. This author uses, for its microchemical recognition, *lead acetate*, which forms a brick-red precipitate with the curcumin, and *sulphuric acid*, which colors it crimson.

d. Coloring Matters dissolved in the Cell-sap.

183. The pigments dissolved in the cell-sap have been little studied. There are usually distinguished only two different pigments or groups of closely-related compounds: namely, *anthocyanin* or cyanin, and *anthochlorin* or xantheïn. The first of these produces red, blue, or blue-green colors, and the latter the yellow and yellow-brown tones.

a. Anthocyanin.

184. Anthocyanin is readily soluble in water, and has in this solution, according as its reaction is more or less acid or alkaline, a red, violet, blue, blue-green, green, or yellow-green color. It is wholly decolorized by strong alkalies. It is also soluble in alcohol and ether.

Whether all the pigments called anthocyanin are chemically identical must be determined by future studies. Hansen (III) believes in the identity of the red and blue pigments prepared from various organs, on the ground of his spectroscopic investigations.

But N. J. C. Müller (I) has recently endeavored to show that a considerable number of very different compounds have heretofore been called anthocyanin. But, as this author gives no account of the mode of preparation of his solutions, which were studied only as to their behavior with caustic potash and sulphuric acid, and with the spectroscope, it remains uncertain how far their different behavior is to be attributed to foreign admixtures.

I may remark here that secretions of pigment of a blue or violet color and of a granular or crystalline structure have been observed in various plants. These consist most probably of a compound of anthocyanin with another substance not yet recognized, perhaps a tannin. These secretions may be finely seen in the epidermis of the petals of various species of *Delphinium*. In the cell from the epidermis of the petal of *D. formosum*, shown in Fig. 29, the beautifully blue secretions, which consist plainly of delicate needles, lie in the violet cell-sap. But in this place the pigment deposits may present the most various forms.

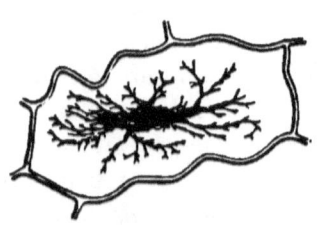

Fig. 29.—Cell of the epidermis of a petal of *Delphinium formosum*.

β. Anthochlorin.

185. The yellow pigments dissolved in the cell-sap are distinguished from xanthin, which occurs in the chromatophores, by never being colored blue by concentrated sulphuric acid, according to Courchet (I, 361-2). In other respects, the yellow pigments contained in the cell-sap of various plants show very different relations to chemical reagents, as Courchet has shown. But our knowledge of them is still too fragmentary to permit their classification.

c. Coloring Matters which are first contained in the Cell-contents, but later penetrate the Wall.

186. According to Sanio (II, 202), Naegeli and Schwendener (I, 501), and Praël (I, 67), all or nearly all the pigments of the dye-woods belong in this category. This follows with greater probability from the study of dried

woods, which usually contain in the cavities of the cells of the medullary rays and of the wood-parenchyma, granules of the same character as the pigments which incrust their walls.

Naegeli and Schwendener have studied from this point of view the madder pigments and berberin, which have, however, recently been placed in other parts of the system of organic compounds (cf. §§ 158 and 212).

187. According to the investigations of A. Rosoll (I, 137), the yellow pigment contained in the bracts of the involucre of various species of *Helichrysum*, which he calls *helichrysin*, belongs to this group. According to Rosoll, it is associated with the protoplasm in the young bracts, and only penetrates the membrane in old cells. Helichrysin has not yet been studied macrochemically, but it is characterized microchemically by being soluble with difficulty in cold water, but readily so in hot water, alcohol, ether, and organic acids. It is insoluble in benzol, chloroform, and carbon bisulphide. *Mineral acids*, as well as *alkalies*, color it a beautiful purple-red. *Lead acetate* precipitates the pigment from its solution with a red color.

188. It seems also probable, from the researches of Naegeli and Schwendener (I, 504), that *anthocyanin* can incrust the cell-walls under some circumstances. At least, the pigment extracted from the seed-coat of *Abrus precatorius* showed a relation to acids and alkalies quite corresponding to that of anthocyanin; while, on the other hand, membranes completely washed out were deeply stained by the cell-sap pressed from red petals of flowers.

f. Coloring Matters which only occur deposited in the Cell-wall.

189. Coloring matters with the above characteristic are widely distributed among the lower plants. Thus two pigments are found among the *Cyanophyceæ*, according to Naegeli and Schwendener (I, 505), glœocapsin and scytonemin, which are completely restricted in their occurrence to the cell-wall, and never occur in the cell-contents; so that it must be supposed that they are formed directly in the wall.

Glœocapsin is red or blue, and becomes red with *hydrochloric acid* (rose-red, reddish orange, or bluish red), and blue or blue-violet with *caustic potash*. It occurs chiefly in *Glœocapsa*.

Scytonemin gives a yellow to dark-brown color to the membranes of *Scytonema* and various other algæ. We have recent studies of its characteristics by Correns (I, 30). According to these, it is especially distinguished by taking a blue-violet color, greatly resembling that which appears in the cellulose reaction, with *chloroiodide of zinc*, or with *iodine* and *sulphuric acid*. Scytonemin is destroyed by *eau de Javelle* (cf. § 12, 4), and threads so treated no longer give the above-mentioned reactions with iodine.

With *acids* scytonemin takes a green color which is restored to the original color on the neutralization of the acid. Even *sulphurous acid* acts in this way and does not destroy the pigment.

Alkalies color it more red-brown. Sheaths which had been treated with caustic potash, although the pigment appeared unchanged after the potash was washed out, no longer took the violet color with chloroiodide of zinc, according to Correns. No macrochemical studies of the composition of these substances have been made.

190. Very numerous and various pigments occur in the membranes of the fungi, according to the investigations of Bachmann (II) and Zopf (V). But, since these have hitherto been studied almost exclusively in their macroscopic relations, and very little has been established as to their chemical composition, the student may be referred to Zopf's account of them (II, 418).

191. The pigments deposited in the membranes of the lichens have lately been studied by Bachmann (IV). He distinguishes by their microchemical behavior sixteen different pigments, whose chief reactions are shown upon the table on the next page.

TABLE TO ACCOMPANY § 191. REACTIONS OF LICHEN PIGMENTS.

Name of the Pigment or of the Lichen bearing it.	Appearance of the Pigment.	KOH.	NH_3.	$Ba(OH)_2$.	HNO_3.	H_2SO_4.	Special Reactions.
Lecidea-green	green				copper- or wine-red		first KOH, then HCl: blue
Aspicilia-green	green						HNO_3: brighter and purer green
Bacidia-green	green				violet	violet	HCl: violet
Thallodima-green	green	violet			indistinctly purple-red		HCl: indistinct purple-red
Rhizoid-green	bluish green	olive-green to brown			olive-green		
Biatora atro-fusca	blue	dissolves with greenish-blue color			violet, then yellow; finally decolorized	dissolves	H_2O: insoluble
Phialopsis rubra	brick-red		colored dirty purple-red		colored violet		
Lecanora-red	purple-red		colored deep violet		brighter colored		
Sagedia declinans	bluish red	blue (green)	first greenish blue, then gray-black	blue			first KOH, then HNO_3, then H_2SO_4: violet crystals
Verrucaria Hoffmanni, f. purpurascens	rose-red	dark green	dark green	dark green			first KOH, then H_2SO_4, then HNO_3: blackish
Bacidia fusco-rubella	yellowish brown	violet	violet	violet			dilute H_2SO_4: bright yellow
Sphæromphale clopismoides	leather-brown	deep olive-green			bright yellow		
Segestria lectissima, perithecia	yellow-brown	auroral red					
Segestria lectissima, entire tissue	brown and colorless						conc. H_2SO_4: deep violet, finally gray
Parmelia glomellifera	leather-brown	dirty brown to olive-brown	blue, then violet, finally gray				$CaCl_2O_2$: first blue, then gray; finally decolorized
Parmelia-brown	yellow to blackish brown				bright red-brown		

g. Coloring Matters which are deposited upon the Cell-wall.

192. In many lichens colored compounds occur which are attached externally to the membranes. They are mostly crystalline, more rarely amorphous.

Amorphous excretions of pigment were found by Bachmann (IV, 27) only in two lichens, and he has named them, from the lichens in which they occur, Arthonia-violet and Urceolaria-red. *Arthonia-violet* occurs in all parts of *Arthonia gregaria* and is especially distinguished by being somewhat soluble in cold, but readily só in hot, water. It is also soluble in *alcohol* with a wine-red color. It is dissolved by a solution of *caustic potash* with a violet color, but is insoluble in lime and baryta waters. It dissolves in *sulphuric acid* with an indigo-blue color, passing later into mallow-red; and in *nitric acid* with a red color.

Urceolaria-red occurs in the thallus of *Urceolaria ocellata*. It is characterized microchemically by not being changed by alcohol, lime-water, or ammonium carbonate. It is dissolved with a greenish-brown color by *caustic potash* solution and *baryta-water*, as well as by concentrated *nitric* and *sulphuric acids*. A solution of *calcium chloride* decolorizes it.

193. The substances already described elsewhere, emodin, chrysophanic acid, and calycin (cf. §§ 140, 141, and 165), belong to the crystalline excretions of the lichen-fungi, as also a series of other so-called lichen acids, which have heretofore been studied almost exclusively macroscopically (cf. Schwarz II and Zopf II, 401).

But I will discuss somewhat more in detail certain recently described fungus-pigments.

α. Thelephoric Acid.

194. Zopf (V, 81) designates as thelephoric acid a pigment extracted from various species of *Thelephora*, whose solutions are of a beautiful red color, while the solid crystalline pigment has a violet-blue or indigo-blue color. This partly forms an incrustation of the membrane, partly a crystalline deposit upon it.

Its behavior with *ammonia*, which gives it a splendid blue color, is especially characteristic of thelephoric acid; while *caustic potash* and soda produce a more bluish-green color. Thelephoric acid is also, according to Zopf, insoluble in water, ether, chloroform, petroleum-ether, carbon bisulphide, and benzol, but is pretty readily dissolved by warm *alcohol*. Concentrated sulphuric or hydrochloric acid neither changes the color nor dissolves the solid pigment; but concentrated *acetic acid* dissolves it with a rose-red or wine-red color, *nitric acid* with a yellow color, and dilute *chromic acid* with a dark chrome-yellow color.

β. Xanthotrametin.

195. Zopf calls by the name xanthotrametin a red pigment which is deposited on the membranes of *Trametes cinnabarina* in the form of granular brick-red incrustations. This dissolves in concentrated *nitric acid* with a deep orangered color, in *hydrochloric acid* with an orange-yellow, and then more reddish, shade, in *sulphuric acid* with a rose-red color, and in glacial *acetic acid* with a yellow color. *Dilute* sulphuric acid dissolves it with an at first orange-yellow color which then becomes redder.

Xanthotrametin is dissolved by *ammonia* and *sodium carbonate* with a yellow color, by dilute *caustic soda* and *lime-water* with a yellow color, becoming paler, and by dilute *caustic potash* with a yellow color which quickly passes into reddish. In ferric chloride it is insoluble.

γ. Pigment of *Agaricus armillatus*.

196. This forms, according to Bachmann (II, 7), crystalline cinnabar-red slivers or lamellæ which are insoluble in alcohol and ether, but dissolve in an aqueous or alcoholic solution of *caustic potash* with a red-violet color which soon goes over into dark yellow.

δ. Pigment of *Paxillus atrotomentosus*.

197. Thörner (I) has extracted a pigment from the above-named fungus which is deposited in crystalline form upon the hyphæ. These crystals are colored brown only at the surface of the fungus, those in its interior being at most pale gray or yellowish. In the air the colorless crystals gradually assume a brown color. According to Thörner, they represent a hydroquinone-like body which gradually passes over into the corresponding quinone. For the recognition of the quinone, Bachmann (II, 7) recommends strongly dilute *caustic potash* or *soda*, which instantly dissolves it with a greenish-brown color.

9. Tannins.

198. All those substances which give a blue-black or green color with iron salts are commonly designated as tannic acids or tannins. There belong here, of the better known compounds, especially:

Pyrocatechin, $C_6H_4(OH)_2$;
Pyrogallic acid, $C_6H_3(OH)_3$;
Protocatechuic acid, $C_6H_3(OH)_2.COOH$;
Gallic acid, $C_6H_2(OH)_3.COOH$;
Gallo-tannic acid, $C_{14}H_{10}O_9$ (= digallic acid?); and besides these there are many other compounds whose constitution is not yet certainly determined (cf. Beilstein, III, 431 ff.). We have no trustworthy methods for the certain microchemical distinction of these substances, although this is the more to be desired since we have certainly to do with very different substances and, as Reinitzer (I) has lately shown, it is very hazardous to assume a common physiological function for this whole group of compounds.

But a detailed account of the methods used for the recognition of the whole group of tannins seems to be demanded by their wide distribution in plants. The following reactions have been made use of in their study:

a. Iron Salts.

199. Of the iron salts, an aqueous solution of ferric chloride was formerly chiefly used; but it has the disadvantage that it has nearly always an acid reaction, and when used in excess with the tannins which give a green color with iron, it very quickly stops the reaction. H. Moeller (I, LXIX) used, however, a solution of anhydrous *ferric chloride* in water-free ether as a reagent for tannin. This is especially adapted to the study of large parts of plants, such as whole leaves or pieces of them.

Loew and Bokorny (I, 370, note) have recently used for the recognition of tannin in algæ a concentrated aqueous solution of *ferrous sulphate*, in which the algæ were allowed to be exposed to the air for from twelve to twenty-four hours. I have obtained in this way a very intense reaction in *Spirogyra* and *Zygnema*. This may be hastened by warming to 60° C.

A very rapid reaction is produced by *ferric acetate*, according to Moeller (I, LXIX), in the form of the concentration of the officinal *tinctura ferri acetici*, which contains about 5% of iron or about 20% of $Fe_2(C_2H_3O_2)_6$.

β. Cupric Acetate.

200. Cupric acetate, which was introduced into microchemistry by Moll (I), has the advantage that it forms an insoluble precipitate with tannins. This is brownish in color, but takes, on subsequent treatment with iron salts, a blue or a green color according to the kind of tannin concerned. Moll places the tissues to be studied in a concentrated aqueous solution of cupric acetate and leaves them in it from eight to ten days, or as much longer as may be desired. Sections prepared from this material are treated for a few minutes with a 5% solution of ferric acetate, and, after it is washed out, may be preserved in glycerine or glycerine-gelatine.

If it is desired at the same time to fix the cell-contents, one may use, according to Klercker (I, 8), a concentrated alcoholic solution of cupric acetate* instead of the aqueous solution. The pieces of tissue should be left several days, at least, in it.

---

* This solution must be kept in the dark.

γ. Potassium Bichromate and Chromic Acid.

201. Most tannins form with *potassium bichromate* a voluminous precipitate which is bright brown or blackish brown according to the quantity of tannin present, and which is insoluble in water or in an excess of the salt, and, according to Moeller (I, LXIX), probably consists of purpurogallin.

Potassium bichromate is most conveniently used by placing large pieces of the tissue to be examined for one or several days in a concentrated aqueous solution of this salt, and then preparing sections after washing out the bichromate. These sections generally show the precipitate in the places where the tannin was formerly present. They may be preserved unchanged in glycerine or glycerine-gelatine. Thicker sections or larger fragments may also be transferred to Canada balsam. But of course they must first be dehydrated with alcohol and cleared with clove-oil or the like (cf. §§ 14–22).

The precipitate produced by potassium bichromate does not change its color with iron salts, or, as Overton (II, 5) has shown, with sulphurous acid. Even hydrogen peroxide does not attack it.

To obtain a rapid reaction J. af Klercker (I, 8) recommends for many cases that the objects be plunged in a boiling solution of potassium bichromate. In fact an immediate reaction may thus be obtained with algæ and sections of higher plants.

According to Moeller (I, LXX), the diffusion of the bichromate is much hastened by the addition of a few drops of acetic acid.

202. Dilute *chromic acid* of about 1% appears to give a reaction similar to that of potassium bicarbonate, and may be used, as well as mixtures of chromic and osmic acids, for the recognition of tannins. These have the advantage of fixing the cell-contents well at the same time.

But it should be noted that, according to Nickel (I, 74), various compounds not related to the tannins give similar precipitates with potassium bichromate.

δ. Osmic Acid.

203. Osmic acid is rapidly reduced by tannic acids and very soon forms with them a solution sometimes bluish and sometimes brownish, and finally a black precipitate. For the reaction a 1% solution should be used. The osmic acid is regenerated and the preparation wholly decolorized by *peroxide of hydrogen.*

If hydrochloric acid be first added to the preparation and then 1% osmic acid, there appears, according to Dufour (I, 3 of separata), in a few minutes a blue color, and soon, if much tannic acid be present, a blue precipitate.

As Overton (II, 5) has shown, albuminoids saturated with tannins are browned; thus he obtained a beautiful brown coloring of protein crystalloids on leaving sections from the endosperm of *Ricinus*, from which the fat had been removed, for about ten minutes in a dilute solution of tannin, and transferring them, after careful washing, to 2% osmic acid.

ε. Ammonium Molybdate.

204. Gardiner (III) used for the recognition of tannins a concentrated solution of ammonium molybdate in a saturated ammonium chloride solution. This gives with most tannins a yellow precipitate, with digallic acid a red one, and with gallic acid a compound soluble in ammonium chloride solution.

ζ. Sodium Tungstate.

205. Brämer lately recommends (I) for the recognition of tannin a solution of 1 gram of sodium tungstate and 2 grams of sodium acetate in 10 ccm. of water. This should give a brown precipitate with gallic acid, and a reddish-yellow one with gallotannic acid. But the presence of tartaric or citric acid hinders the reaction.

η. Alkaline Carbonates.

206. Alkaline carbonates cause the precipitation, in cells containing tannin, of globular or rod-shaped bodies which, when freshly precipitated, may be redissolved on washing out the carbonate. But after a time they lose this capacity and are gradually transformed into solid and brittle bodies. The same precipitates are formed when tannin and ammonium carbonate are brought together in a test-tube; but if the carbonate solution be very dilute, it only occurs when a gradual accumulation of this salt is made possible; which Klercker (I, 42) accomplished by placing the tannin solution in a capillary tube sealed at one end, and then putting the whole in a large vessel filled with the carbonate solution. Since, according to Klercker's researches, these precipitates have been seen only in cells containing tannin and always give the tannin reactions with potassium bichromate and other reagents, it would appear that the alkaline carbonates may well be used as reagents for tannin.

But not all tannins are precipitated by them; for instance, gallic acid is not.

207. In order to carry out the reaction described, sections of the plants to be studied are placed on the slide in a drop of a 1-5% solution of an alkaline carbonate or the plants are cultivated for some time in a very dilute, about .02%, solution. The carbonates of potassium, sodium, and ammonium seem to be equally active, but the ammonium salt has been most used. *Ammonium chloride* seems also to act in the same way; but the bicarbonates produce no precipitate.

A very suitable object for study is found in the stem or petiole of *Ricinus communis*, which contain red pigment-cells in the pith and bark. If longitudinal sections of these are placed in a 1% ammonium carbonate solution, granular precipitates appear in a short time, which gradually collect together and, after about an hour, have drawn almost the entire pigment to themselves (cf. Fig. 30). *Spirogyra* cells which contain tannin also form excellent objects for study, since a precipitate is almost immediately formed in them by ammonium carbonate.

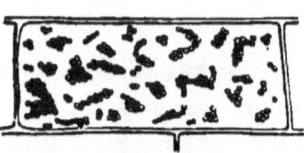

FIG. 30.—Cells from the pith of the petiole of *Ricinus communis*. The section had lain an hour in a 1% solution of ammonium carbonate.

9. Live-staining with Methylene Blue.

208. As was shown by Pfeffer (II, 186), methylene blue is accumulated by tannin-bearing cells. In a solution of this pigment the cell-sap which contains tannin first takes an evidently blue color, and then there usually occur within these cells deep-blue precipitates of different forms, which consist of a compound of tannin and methylene blue and may finally remove all pigment from the cell-sap.

This reaction seems to take place in all tannin-bearing cells, and is especially valuable because it can be conducted on the living cells, and without the diminution of their vitality. For its application a solution should be used which contains one part of methylene blue in 500,000 parts of filtered rain-water. In this the tissues are left for from one to twenty-four hours. Of course a large quantity of fluid must be used to allow an abundant accumulation of coloring matter.

But methylene blue is accumulated by other substances than tannins; according to Waage (I, 253), by phloroglucin.

10. Alkaloids.

209. Under the name alkaloids is included a large group of natural basic compounds which contain nitrogen and show a certain agreement in many chemical reactions.

They are all precipitated by *phospho-molybdic acid*, which may conveniently be used in a 10% aqueous solution. But this reaction cannot be generally used in microchemistry, since this acid precipitates many other compounds, e.g. the proteids. But, according to Errera (V), these may be distinguished from the alkaloids by extracting the alkaloids from a part of the sections to be studied before treatment with phospho-molybdic acid, while the proteids are completely precipitated. Errera found especially adapted for this use alcohol which contains in each 20 ccm. a gram of crystallized tartaric acid. He allows this to act upon the sections from half an hour to twenty-four hours, when a complete solution of the alkaloids is effected, with the simultaneous precipitation of the proteids. If no precipitation takes place in the sections so treated on the addition of phospho-molybdic acid, while it does occur in the untreated sections, one may deduce the presence of alkaloids in the objects concerned, provided that these were the only substances soluble in tartaric-acid-alcohol which are precipitated by phospho-molybdic acid. But this is by no means the case (cf. Molisch, I, 15), and the above reactions can give reliable results only in very special cases.

The same is true for the other group-reactions for the alkaloids; but the value of the special reactions for the different alkaloids, to which we now pass, is so much the greater.

a. Aconitine, $C_{33}H_{43}NO_{12}$.

210. According to Errera, Maistriau and Clautriau (I) a solution of *iodine and potassium iodide* is best adapted for the microchemical recognition of aconitine, with which it forms a red-brown precipitate. These authors also obtained a carmine-red color in presence of aconitine with *sulphuric acid* diluted with $\frac{1}{4}$ or $\frac{1}{2}$ of its volume of water, especially after the moistening of the preparations with a solution of cane-sugar.

b. Atropine, $C_{17}H_{23}NO_3$.

211. De Wevre (I) used a solution of *iodine and iodide of potassium* for the microchemical recognition of atropine, in cells containing which a brown precipitate is produced. After some time star-shaped crystallizations with a metallic lustre appeared. Phospho-molybdic acid, which gives a yellow precipitate, is also applicable in many cases.

c. Berberin, $C_{20}H_{17}NO_4 + 4\frac{1}{2}H_2O$.

212. Berberin occurs, according to the consonant statements of Naegeli and Schwendener (I, 503), O. Herrmann (I, 14), and Rosoll (I, 20), in the young cells of *Berberis vulgaris* as a golden-yellow fluid in the cell-cavity; while in the older parts it incrusts especially the cell-walls of the xylem.

For its microchemical recognition O. Herrmann treats the sections first with alcohol and then adds dilute *nitric acid* (one part of acid to about fifty parts of water). The golden-yellow color passes over at once into brownish yellow, and soon golden-yellow star-shaped crystals are deposited, while the cell-sap becomes gradually colorless. The same crystals of berberin nitrate may be obtained by placing the sections directly in dilute nitric acid.

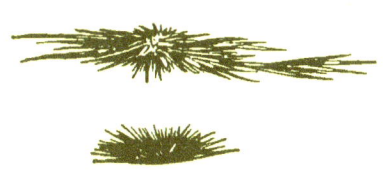

FIG. 31.—Groups of crystals of berberin nitrate obtained by placing longitudinal sections of the bark of *Berberis vulgaris* in a 2% solution of nitric acid.

I use 2 parts of the officinal nitric acid in 100 parts of water. The berberin nitrate separates out in the form of clustered crystals (cf. Fig. 31) in the interior of the berberin-bearing cells. These crystals are strongly doubly refractive and so far pleochroic (§ 356, note) that they appear quite colorless in a certain position of the nicols, while after rotation through 90° they appear deep golden yellow.

Herrmann also used *ammonium sulphide*, which gives a brownish color with it, for the recognition of berberin.

Rosoll used a solution of *iodine and potassium iodide* for the same purpose. This is added to the sections in small quantity after preliminary treatment with alcohol. There are then formed very characteristic hair-like crystals, arranged in tufts, which are green or, in the presence of large quantities of the reagent, yellowish or reddish brown, and are soluble in sodium hyposulphite. The crystals thus obtained cannot be preserved in Canada balsam, according to my experiments.

If *hydrochloric acid* be added to the yellow aqueous extract of the dried bark, there are formed, according to Naegeli and Schwendener (I, 504), numerous yellow and often radially grouped needle-like crystals of berberin chloride.

d. Brucine, $C_{22}H_{26}N_2O_4 + 4H_2O$.

213. For the recognition of brucine, which accompanies strychnine in the seeds of *Strychnos nux-vomica*, Lindt (II, 239) used a mixture of five drops of *selenic acid*, of specific gravity 1.4, and one or two drops of *nitric acid*, of specific gravity 1.2. He allowed this to run under the cover-glass to thin sections whose fat had been extracted by petroleum-ether. The stratified cell-walls then quickly became bright-red, but gradually changed to orange and yellow, while the cell-contents remained colorless. Lindt concluded from this that only the walls contain brucine (cf. also § 219).

e. Colchicine, $C_{22}H_{25}NO_6$.

214. Colchicine occurs, according to O. Herrmann (I, 18), in the corm of *Colchicum autumnale*, in the contents of two or three rows of cells which immediately surround the vascular bundle and are distinguished from the neighboring parenchyma-cells by being free of starch. For the recognition of colchicine this author used *ammonia*, which turns it a deep yellow.

Errera, Maistriau and Clautriau (I) have lately characterized it as giving a yellow color with *sulphuric acid* diluted with two or three volumes of water, a brown-violet with *sulphuric* and *nitric acids*, a brown with *iodine*, and a yellowish precipitate with *potassic-mercuric iodide* and *hydrochloric acid*.

f. Corydalin, $C_{18}H_{19}NO_4$.

215. According to Zopf (VI, 113), corydalin shows the following reactions: It is precipitated from aqueous solutions of its salts by caustic *alkalies* and alkaline carbonates, but is redissolved in an excess of the former. A brown precipitate is produced in solutions of its salts by solutions of *iodine* or of *iodine and potassium iodide*, a yellow one by *potassium chromate*, a white one by *mercuric chloride*, a yellow one by *gold* and *platinum chlorides* and by *sodium metatungstate*, a yellowish-white one by *potassic-mercuric* *chloride*, a white one by *potassium sulpho-cyanide*. The solution in spirit is precipitated by strong alcohol.

For its microchemical recognition Zopf used especially *ammonia*, which produces a dark gray, granular precipitate in the cells containing corydalin, and *iodine with iodide of potassium*, which causes a deep red-brown precipitate. *Picric acid* also causes, according to Zopf, a readily recognizable precipitate in corydalin-bearing cells.

g. Cytisine, $C_{20}H_{27}N_3O$.

216. Cytisine occurs, according to Rosoll (I, 24), in all the organs of *Cytisus Laburnum*, but most abundantly in the contents of the cells of the ripe seed. This author uses the following reactions for its recognition:

Iodine and iodide of potassium give, even when very dilute, a brownish color and then a dark red-brown precipitate, soluble in sodium hyposulphite.

Picric acid, added to thin sections, produces in a short time scale-like crystal-groups of a reddish-yellow color.

Concentrated sulphuric acid dissolves cytisine with a bright reddish-yellow color; if very small bits of solid *potassium bichromate* be added to this solution, it becomes first yellow, then brown, and finally green.

Phospho-molybdic acid produces immediately a yellow turbidity in the sections.

h. Opium Alkaloids.

217. According to the investigations of Clautriau (I), several of the so-called opium alkaloids occur in the latex of the living plant in *Papaver somniferum*.

1. Morphine, $C_{17}H_{19}NO_3$.

The presence of morphine is recognized by this author by the fact that the latex gives precipitates with *iodine and potassium iodide, potassium-bismuth iodide, potassium-calcium iodide, potassic-mercuric iodide*, and *phospho-molybdic acid;*

that it reduces *iodic acid;* and that it is colored red-brown by a 2% solution of *titanic acid* in sulphuric acid, or deep violet by a solution containing five drops of *methylal* ($CH_2(OCH_3)_2$) to a ccm. of concentrated sulphuric acid.

2. Narcotine, $C_{22}H_{23}NO_7$.

Clautriau deduces the presence of narcotine in the latex from its becoming colored reddish orange with a solution of *sodium selenate* in sulphuric acid, as also happens with mixtures of morphine and narcotine. Besides, it gives a precipitate with *palladous chloride* and with *iridous chloride*, while morphine and codeine give no precipitate with the former, and but a slight one with the latter reagent.

3. Narceïne, $C_{23}H_{29}NO_9$.

According to Clautriau, a yellow color following the addition of the solution of *methylal*, mentioned under "Morphine," indicates the presence of narceïne in the latex.

i. Piperine, $C_{17}H_{19}NO_3$.

217a. Piperine, which has been recognized in the fruits of various *Piperaceæ*, is slightly soluble in boiling water, more easily soluble in alcohol than in ether, readily soluble in *benzol*, insoluble in dilute acids.

For its microchemical recognition concentrated *sulphuric acid* may be used. This dissolves piperine with a yellow color which later becomes dark brown, and, after 20 hours, greenish brown (Husemann I, 491). But, since various other substances give the same reaction, it can only be of value as negative evidence through its failure to appear.

We owe to Molisch (I, 27) a very useful method of recognizing piperine microchemically. A drop of alcohol is first placed on the sections on the slide and dissolves the piperine. Then the sections are covered with a cover-glass, and the alcohol is allowed to evaporate until it occupies only about a quarter of the space under the cover. Then water is added and causes at once a strongly milky turbidity, if sections of the pepper-fruit are used. This turbidity is due chiefly to the yellow resin which is also dissolved in the alcohol. After a time (about a quarter of an hour) there are seen characteristic colorless crystals, especially abundant at the edge of the cover-glass, which have very commonly an approximately sabre-shaped outline, but are not seldom grown together in most various ways, as is shown in Fig. 32, which represents crystals obtained by the method described. All these crystals give the piperine reactions above described and undoubtedly consist of that substance.

FIG. 32.—Crystals of piperine.

Similar piperine crystals are formed, according to Molisch (I, 28), within or in the vicinity of the piperine-cells of pepper, if thin sections are placed under a cover-glass in water or glycerine and kept in a moist chamber for several hours. "Sections which are pressed and rubbed under a cover-glass in water show piperine crystals within the first quarter of an hour."

k. *Sinapine*, $C_{16}H_{23}NO_5$.

218. This occurs in the seeds of white mustard as sinapine sulphocyanide ($C_{16}H_{23}NO_5.HCNS$).

According to Molisch (I, 31), it is best recognized with a concentrated solution of *caustic potash*. Sections of white mustard seeds placed in this become at once yellow, and, on warming, deep orange. But this reaction has a very limited value, since the glucoside sinalbine, also found in white mustard-seeds, becomes yellow with caustic potash.

l. *Strychnine*, $C_{21}H_{22}N_2O_2$.

219. Strychnine is dissolved without color by a concentrated *sulphuric acid;* but if solid *potassium bichromate* be added to the solution, a beautiful violet color appears at once. This reaction has been used by Rosoll (I, 17) for the microchemical recognition of strychnine. He placed thin sections from the seeds of *Strychnos nux-vomica* first in concentrated sulphuric acid, and observed that the contents of the endosperm-cells become plainly rose-red from the presence of proteids and sugar (cf. § 227), except the oil-drops, which remained uncolored. If now a small fragment of potassium bichromate be added, the previously colorless oil-drops take a beautiful violet color. Rosoll concludes from this that the strychnine is dissolved in the oil-drops.

Lindt (II) used a solution of an excess of *ceric sulphate* in concentrated sulphuric acid for the recognition of strychnine. In spite of the gradual reddening of the ceric oxide, this reagent remains fit for use a long time. Before its use the sections should be treated with petroleum-ether and alcohol for the removal of fatty oils, grape-sugar, and brucine. If the solution be then added to the sections, all the cell-walls become at once colored more or less strongly violet-blue, while the contents of the cells at first remain colorless. But after some time the reaction is disturbed by various circumstances. Lindt concludes from this observation that the strychnine is contained in the cell-walls. But Rosoll (I, 18) expresses the opinion that, during the extraction with petroleum-ether, the strychnine, formerly dissolved in the fatty oil, is removed with it and becomes partly diffused into the cell-walls.

m. Theobromine, Dimethyl-xanthin, $C_8H_8N_4O_2$.

220. The alkaloid contained in the cocoa-bean, theobromine, is best recognized, according to Molisch (I, 23), by means of *gold chloride*, by adding to sections upon a slide, first a drop of concentrated hydrochloric acid and then, after about a minute, a drop of a 3% solution of gold chloride. As soon as a part of the fluid has evaporated, long yellow needles are formed at the edge of the drop, and finally unite into feathery or tufted groups. The crystals consist of the double chloride of gold and theobromine, $C_7H_8N_4O_2.HCl.AuCl_3$.

n. Caffeine, Theine, Methyl-theobromine, Trimethyl-xanthin, $C_8H_{10}N_4O_2 + H_2O$.

221. Molisch recommends (I, 7) *gold chloride* for the microchemical recognition of caffeine. It is used in the same way as for theobromine, and there are formed tufted radiating needles of the composition $C_8H_{10}N_4O_2.HCl.AuCl_3$, which cannot certainly be distinguished from those of the corresponding theobromine compound.

Hanausek (I) has lately called attention to the fact that, especially when the gold chloride solution contains more than 3% of the salt, a drop of it with concentrated hydrochloric acid will, in any event, form yellow crystals on evaporation. But these are distinguished from the caffeine-gold chloride crystals by the fact that they never form sharp-pointed or tufted needles, like those of the caffeine-gold compound, but consist partly of very short crystals arranged in zigzag, and partly of very long, delicate, yellow, rod-like prisms and of plates with rectangular projections.

Molisch gives also another method for recognizing caffeine. According to this, sections are warmed on the slide in a drop of distilled water until it bubbles, then allowed to dry at the ordinary temperature, and the residue is taken up with a drop of *benzol*. On the evaporation of the benzol, the caffeine is deposited at the edge of the drop in the form of numerous colorless needles.

o. Veratrine, $C_{37}H_{63}NO_{11}$.

222. Veratrine occurs, according to Borscow (I, 58), in the subterranean parts of *Veratrum album*, especially in the walls of the cells of the epidermis and of the bundle-sheath (endodermis). This author used for the recognition of veratrine a mixture of one part of concentrated *sulphuric acid* with two parts of water, in which the sections to be examined are directly placed. The parts containing veratrine then become first yellow, then reddish orange, and finally red-violet.

p. Xanthine, $C_5H_4N_4O_2$.

222a. Xanthine has been recognized by Belzung (II, 49) in considerable quantity in the seedlings of *Cicer arietinum*. It crystallizes out in the interior of the cells of plants placed in glycerine.

11. Nitrogenous Bases.

Nicotine, $C_{10}H_{14}N_2$.

223. The following reactions have been recommended for the recognition of nicotine by Errera, Maistriau and Clautriau (I): *Phospho-tungstic acid* gives a heavy precipitate, at first yellow, then yellowish green; *mercuric chloride*, a white one, soluble in an excess of ammonium chloride with heat; *potassic-mercuric iodide*, an abundant white one; *platinum chloride*, a yellowish-white one, soluble at 70° C.; *iodine and potassium iodide*, a brownish-yellow one which disappears later.

12. The Proteids and Related Compounds.

224. Although the albuminoid substances or proteids undoubtedly belong to the most important constituents of the plasma-body of the cell, we still know very little of their chemical constitution. But, at all events, we have here to do with a group of dissimilar compounds, and already various investigators have distinguished a large number of different proteids. But since no classification of the proteids has yet found general acceptance, and especially since the exact microchemical distinction of the proteids which are recognized macrochemically is not yet possible, it seems best not to discuss these researches here; and so much the more so since they have not led to any morphologically or physiologically interesting results, so far as concerns the vegetable organism. But it seems important to briefly collate the methods heretofore used for the microscopical recognition of the proteids in general, and then to describe a few related substances which have been microscopically recognized in the protoplasm.

a. Reactions of the Proteids.

α. Solutions of Iodine.

224a. With iodine the proteids take a yellow or brown color, according to the strength of the solution, and this reaction has been much used for their recognition, although many other substances give the same reaction. But in many cases the behavior of a doubtful body with iodine may give a clew to its nature. It is best to use a solution of iodine and potassium iodide containing more iodine than is used for the recognition of starch, since iodine is taken up much less freely by proteids than by starch.

β. Nitric Acid.

225. Ordinary concentrated nitric acid gives a yellow color with proteids by the formation of the so-called *xanthoproteic acid*. This reaction may be hastened by gentle warming. The color becomes considerably deeper by the addition of caustic potash or ammonia, since the xanthoproteic salts of potassium and ammonium are more deeply colored than the free acid. But this reaction is not entirely trustworthy, since not only do tyrosin and various oxy-aromatic compounds give the same reaction, but also certain oils, resins, and alkaloids (cf. Nickel I, 17).

γ. Millon's Reagent.

226. The so-called Millon's reagent is a mixture of *mercuric and mercurous nitrates* and *nitrous acid*. It is best prepared, according to Plugge (I), by dissolving one part by weight of mercury in two parts of nitric acid of specific gravity 1.42, and then diluting it with twice its volume of water. According to Nickel (I, 7), it may be prepared by dissolving 1 ccm. of mercury in 9 ccm. of concentrated nitric acid of specific gravity 1.52 and diluting the solution with an equal volume of water. The reagent so obtained becomes decomposed in time, and may then be restored, according to Krasser (I, 140), by the addition of a few drops of a solution of potassium nitrite.

Millon's reagent gives with proteids a brick-red, or more rose-red, color, which appears usually after some time in the cold, but much more quickly on gentle warming, without any solution of the proteids. The protein crystalloids in the endosperm of *Ricinus*, for example, retain their form unchanged even on heating nearly to the boiling point. This reaction is also very delicate, but it has the disadvantage that it takes place also with a large number of other compounds, in the same way; according to Plugge (I), with all those that contain an OH-group on the benzol nucleus.

227. Reactions similar to that of Millon's reagent are given by *Hofmann's reagent*, which is a solution of mercuric nitrate with traces of nitrous acid, and by the so-called *Plugge's reagent*, which consists of mercurous nitrate with traces of nitrous acid (cf. Nickel I, 12 and 13).

δ. Raspail's Reagent.

227a. The so-called Raspail's reagent consists of a concentrated aqueous solution of *cane-sugar* and concentrated *sulphuric acid*, which are added to the objects to be tested at the same time. Proteids are colored by this rosy-red or somewhat violet. But the reaction does not succeed with all proteids and does occur with various other substances. Many glucosides and alkaloids are colored red by sulphuric acid alone (cf. Nickel I, 37 ff.).

ε. Copper Sulphate and Caustic Potash.

228. Most proteids give a violet reaction on treatment with copper sulphate and caustic potash; but the reaction is neither always very sharp nor, when it succeeds, positive proof of the presence of proteids. The reaction may be conducted in the same way as for cane-sugar (cf. § 121).

This reaction was used on *Spirogyra* by Loew and Bokorny (I, 194) by first placing the plant in a .2% caustic potash solution for half an hour to an hour, then, after washing, in a 10% solution of copper sulphate for an hour, and finally, after repeated washing on the slide, moistening with a 2 % caustic potash solution.

ζ. Alloxan, Mesoxalylurea, $C_4H_2N_2O_4$.

229. Alloxan has lately been proposed by Krasser (I, 18) as a reagent for proteids. It is added, preferably to previously dried sections, in concentrated aqueous or alcoholic solution. It then colors most proteids purple-red, and, according to Krasser, this color is not changed by concentrated caustic soda. But the reaction is hindered by free acids. Of the other organic substances which Krasser has treated in the same way, unfortunately without enumerating them in his publication, only tyrosin, asparagin, and aspartic acid gave the same reaction; and it therefore seems to be caused by the molecular group common to these substances, $CH_2.CHNH_2.COOH$.

But, as Klebs has shown (IV, 699), the value of this reaction is very slight; for various inorganic compounds like phosphates and bicarbonates of the alkalies give a very deep red color with alloxan, and alloxan itself turns red on evaporation in the air; and this color, as well as that produced with proteids, is changed to violet by caustic soda, according to Klebs.

η. Aldehydes.

229a. Various aldehydes, especially *salicylic aldehyde*, $C_6H_4OH.CHO$, *anisic aldehyde*, $C_6H_4.OCH_3.CHO$, *vanillin*, $C_6H_3.OH.OCH_3.CHO$, and *cinnamic aldehyde*, $C_6H_5.CH : CH.CHO$, have lately been recommended for the microchemical recognition of albuminoids by Reichl and Mikosch (I, 34). The objects to be tested are first left for 24 hours in a ½–1% alcoholic solution of the aldehyde used, and then placed on the slide in a mixture of equal volumes of water and sulphuric acid, to which a few drops of ferric sulphate

($Fe_2(SO_4)_3$) have been added. Most proteids then show, either at once or after some time, a coloration depending not only on the nature of the aldehyde used, but also varying with the different albuminoids, and not appearing at all, with many. The colors with the salicylic and anisic aldehydes and vanillin vary between red, violet, and blue; while the cinnamic aldehyde produces yellow or orange-yellow colors with proteids.

Salicylic aldehyde has, according to the statements of the authors mentioned, the advantage of more completely fixing the proteids and of making them more resistent to the dissolving power of sulphuric acid.

θ. Yellow prussiate and ferric chloride.

230. Zacharias (I, 211) has lately reintroduced into microchemistry a method of recognizing proteids first used by Th. Hartig, which may be here referred to, although it is more a staining process than a chemical reaction. This test was carried out by Zacharias by placing the tissues to be tested for an hour in a mixture of one part of a 10% aqueous solution of *potassium ferrocyanide*, one part of water and one part of acetic acid of specific gravity 1.063. This mixture, which must always be freshly prepared, on account of its ready decomposition, is now washed out with 60% alcohol until the washing fluid no longer gives an acid reaction or a blue color with ferric chloride. Then a dilute solution of *ferric chloride* is added, which causes a deep blue coloring of the albuminoids (Berlin blue) in consequence of the ferrocyanide retained by them.

231. To recognize albumen in the cytoplasm of *Spirogyra*, Loew and Bokorny (I) first left the plants an hour in a .1% solution of ammonia, then placed them for 12 hours in a 10% solution of ferrocyanide containing 5% of acetic acid, then washed in cold water, and finally let them remain for 12 hours in a not too dilute solution of ferric chloride. Certain differentiations of the cytoplasm then showed an evident blue color.

t. Pepsin and Pancreatin.

232. Recently, the ferments secreted by the stomach and pancreas, which have the power of converting proteids into soluble compounds (peptones), have been used for their microchemical recognition. Both ferments can now be obtained in very stable form, as pepsin-glycerine and pancreatin-glycerine, of Dr. G. Grübler, Leipzig.

233. Digestion with pepsin may be accomplished by keeping the objects in a mixture of one part pepsin-glycerine and three parts water, acidified with .2% of its weight of chemically pure hydrochloric acid, for an hour, at a temperature of 40° C. The effect of hydrochloric acid may be observed in control-experiments containing the acid alone.

234. Pancreatin-glycerine may be diluted with three times its volume of water and then used in the same way.

The previous treatment of the objects has an important influence upon the reaction. The solution of the proteids takes place most easily in sections taken directly from the living plant. But, in general, alcoholic material is to be preferred, since the digestibility of the substances soluble in water may thus be determined, and clearer images are usually obtained. But the alcohol should act for as short a time as possible (24 hours), since it may influence the digestibility by its longer action.

b. Nucleins.

235. Nucleins have been prepared especially from yeast, from the thymus gland of the calf, from the yolk of eggs, and from salmon-sperm. These are distinguished from proteids especially by the fact that they contain phosphorus. In other respects the analyses of nucleins from different sources show considerable differences. Altmann (II) has lately isolated from these nucleins bodies of uniform composition which he calls *nucleic acids*. These contain about 9% of phosphorus and are quite free from sulphur when pure. They are precipitated by albumen, and Altmann regards the nuclein preparations of various authors as compounds of nucleic acids with various amounts of albumen.

[Malfatti (II) has prepared an artificial nuclein from syntonin and metaphosphoric acid, which yields nucleic acids when treated by Altmann's method.]

236. Zacharias has recently (I and II) attempted to recognize microchemically the general distribution of nucleins, especially in cell-nuclei. He gives as a characteristic reaction for them, their insolubility in *pepsin* and hydrochloric acid (cf. § 233), in which, as well as in .2–.3% hydrochloric acid, they take a sharply defined and peculiarly glistening appearance. But the nucleins are, according to Zacharias, soluble in a 10% solution of *common salt*, in a concentrated solution of *sodium carbonate*, in dilute *caustic potash* solution, in concentrated *hydrochloric acid*, and in a mixture of four parts concentrated hydrochloric acid with three parts water. Further investigations must show how far the macrochemically prepared nucleins and those recognized by the above reactions correspond with each other.

[According to Malfatti (I) and Zacharias (V), the nucleins seem to constitute the so-called chromatin-bodies of the nucleus (cf. § 239).]

c. Plastin

237. Reinke (I) prepared a nitrogenous compound from the plasmodia of *Æthalium septicum*, which he calls plastin. According to Zacharias (I and II), this compound represents the fundamental substance of the cytoplasm and agrees in its microchemical behavior with nuclein, in not being attacked by *pepsin* with hydrochloric acid and in being soluble in concentrated hydrochloric acid. But plastin differs from nuclein in not swelling in a 10% solution of salt after treatment with pepsin, and in being less readily soluble in alkalies and insoluble in a mixture of four parts of pure concentrated hydrochloric acid and three parts water.

It remains to be learned whether plastin is really a single compound or includes a group of related compounds. According to Loew (I), the plastin prepared macrochemically by Reinke is to be regarded simply as an impure albuminoid preparation. This author also showed that the cytoplasm regarded by other authors as free from albumen gives the protein-reaction after being first treated for a time with caustic potash (cf. § 228 and Loew, II).

d. Cytoplastin, Chloroplastin, Metaxin, Pyrenin, Amphipyrenin, Chromatin, Linin, and Paralinin.

238. According to the investigations of Schwarz (I), the protoplasmic body is made up of eight different proteids which are limited in their distribution to very special differentiations of the protoplasm, and which should be comparatively easy to distinguish microchemically according to the tables of their reactions prepared by this author. But if one examines the separate results of his observations, as described in detail, it is found that the most of the reagents used have given very different results even with the few objects examined, and that the author's own observations do not at all always correspond with the special statements of his tables. There can no longer be any doubt that the substances distinguished by Fr. Schwarz do not represent uniform, chemically definable substances. Further studies must show whether the reagents used by him are capable of rendering good service in the study of the morphological elements of the protoplasm. This seems most probable in case of the "pretty concentrated" solution of *copper sulphate* used by Schwarz (I, 116), which dissolves only the chromatin * in the nucleus and fixes all its other constituents well. A mixture of one volume of a 10% aqueous solution of *potassium ferrocyanide*, two volumes of water, and half a volume of glacial acetic acid also dissolves only chromatin; but this reagent is not adapted to fixing, since it causes swelling.

239. But, since the nomenclature introduced by Fr.

---

* [Malfatti (I) states that copper sulphate does not dissolve chromatin, but forms an unstainable precipitate with it.]

Schwarz has been used elsewhere in the literature, at least the names of his eight compounds may be given here.

Two of them occur in the chlorophyll granules, chloroplastin and metaxin. The first of these represents the green fibrillæ within the chloroplasts, and between them is the water-soluble metaxin.

In the nucleus Schwarz distinguished five substances, amphipyrenin, which forms the nuclear membrane; pyrenin, the substance of the nucleoli; chromatin, the strongly staining material of the nuclear framework; and linin and paralinin, the former of which forms a fibrillar network in the nucleus, while the latter fills the meshes of this net.

In the cytoplasm Schwarz finds only one proteid, which he calls cytoplastin.

13. Ferments.

240. It was stated by Wiesner (IV) that pepsin, diastase, and the gum-ferment described by him give characteristic color-reactions with *orcin* and hydrochloric acid, which are microchemically applicable. But Reinitzer (II) showed that these reactions occur with various carbohydrates, and most probably depend upon the fact that the reagent splits off from them furfurol or related compounds.

But Guignard has recently tried to determine the location of emulsin and myrosin, partly by the use of orcin.

a. Emulsin.

240a. Emulsin splits the glucoside *amygdalin*, contained in bitter almonds, into prussic acid, oil of bitter almonds, and sugar. It occurs, according to Guignard (III), in the leaves of *Prunus Lauro-cerasus*, exclusively in a parenchymatous sheath surrounding the vascular bundles. This author reaches this conclusion from the fact that only these cells form prussic acid with a solution of amygdalin; while the spongy and palisade parenchyma, which, on the other hand, forms prussic acid with a solution of emulsin, is plainly to be regarded as the seat of amygdalin.

The contents of the cells containing emulsin become red with *Millon's reagent*, and violet with *copper sulphate and caustic potash*. It seems probable that these reactions are due to the emulsin, since the corresponding cells of the emulsin-free leaves of *Cerasus lusitanica* do not give them.

b. *Myrosin.*

241. For the microchemical recognition of myrosin, which is contained in many *Cruciferæ*, Guignard recommends (II) concentrated *hydrochloric acid* which contains a drop of a 10% aqueous solution of *orcin* in each ccm. If the sections are heated in this solution to near 100° C., a violet color appears in the cells containing myrosin.

In the seeds of black mustard and in other parts of various *Cruciferæ*, this reaction occurs in specialized cells rich in proteids, which alone, as Guignard has experimentally shown, are able to decompose potassium myronate (cf. § 156).

[Spatzier (I) finds myrosin also in the *Resedaceæ* and in the seeds of the *Violaceæ* and *Tropæolaceæ*. Where it occurs in vegetative organs he finds it in a dissolved condition; but in dry seeds it is in the form of solid homogeneous granules of about the size of aleurone grains.]

Part Third.

METHODS FOR THE INVESTIGATION OF THE CELL-WALL AND OF THE VARIOUS CELL-CONTENTS.

A. The Cell-wall.

242. SINCE vegetable cell-membranes belong to the class of absorbent bodies and, as their osmotic relations show, can readily give passage to the most various substances which are soluble in water, it may be assumed that they never consist simply of cellulose and water, in the living plant. Rather, they are always incrusted, within the plant, with a greater or less amount of foreign substances, according to their age and position. How far the varying chemical and physical relations of vegetable cell-membranes are to be attributed to such incrustations of organic or inorganic nature cannot at present be certainly determined.

But it is well established by modern researches that pure *cellulose* is much less than the other constituents in many membranes or parts of membranes. Indeed it is very probable that cellulose is entirely wanting in many parts of membranes.

Positive conclusions concerning these questions can only be reached when we have obtained, by macroscopic researches, a sure means of distinguishing the different constituents of the cell-wall. But our knowledge in this respect is still too fragmentary to make possible any uniform system of the constituents of the cell-wall, although the macrochemical study of the cell-wall has been taken up in various aspects in recent years.

243. I prefer, then, to discuss first in the following pages the kinds of membranes which may primarily be distinguished by their microchemical relations. For some time there have been pretty generally distinguished the pure cellulose wall, the lignified wall, the cuticularized or suberized wall, the gelatinized wall, and fungus-cellulose. In connection with the gelatinized wall may be discussed the remaining plant mucilages and gums and the jelly-formation of the *Conjugatæ*. As related to these various modifications of the membrane may be mentioned the paragalactan-like substances which serve as reserve-materials, callose, and the pectins. Finally, this chapter may describe the preparation of ash and siliceous skeletons, and some methods which have been used in the investigation of the development and finer structure of the cell-walls.

1. The Cellulose Wall.

244. Cellulose is a carbohydrate whose empirical composition corresponds to the formula $C_6H_{10}O_5$. It is especially characterized by its solubility in *cuprammonia* and in concentrated *sulphuric acid*, by its blue or violet color with *iodine and sulphuric acid* or with *chloroïodide of zinc*, and by the fact that from its hydrolytic splitting with sulphuric acid there finally results a fermentable sugar (glucose).

245. But it should be remarked that there are very probably different substances which give the same reactions, and which perhaps represent nearly related isomeric compounds. Thus, according to W. Hoffmeister (I and II), cellulose shows very varying relations, especially toward a 1–5% *caustic soda* solution, in which it is partly soluble, partly insoluble. But an exact microchemical distinction of the kinds of cellulose has not yet been possible, and it is therefore best for the present to call all membranes or parts of membranes which show the above reactions *pure cellulose* membranes, any ash constituents being, of course, quite disregarded.

246. For the microchemical recognition of the cellulose membranes the following reactions are used:

1. Solubility in concentrated *sulphuric acid*. This begins with a strong swelling, which finally passes into complete solution.

2. Solubility in *cuprammonia*. This reagent, which is also known as *Schweizer's reagent*, may be prepared by precipitating cupric oxyhydrate from a solution of cupric sulphate with a dilute solution of caustic soda, then washing the precipitate with water by repeated decantation and filtering, and finally dissolving it in the most concentrated ammonia-water.

A very good reagent may be more simply prepared by pouring 13–16% ammonia-water over copper turnings and letting the whole stand in an open bottle (cf. Behrens II, 55).

Cuprammonia can be preserved for only a limited time. To test its fitness for use, one may use cotton, which it should completely dissolve at once.

3. The blue color with *iodine and sulphuric acid*. According to Russow, this reaction may be best conducted by treating the sections first with an aqueous solution of $\frac{1}{8}$% iodine and $1\frac{1}{8}$% potassium iodide, and then adding a mixture of two parts concentrated sulphuric acid and one part water.

4. The violet color with *chloroïodide of zinc*. This reagent is usually prepared by dissolving an excess of zinc in pure hydrochloric acid and then evaporating the solution to the density of sulphuric acid in the presence of an excess of metallic zinc; the solution is then saturated with potassium iodide and finally with iodine. The chloroïodide may be more simply prepared by dissolving 25 parts of zinc chloride and 8 parts of potassium iodide in 8.5 parts of water, and then adding as much iodine as will dissolve (Behrens II, 54). [I have used with excellent results a preparation obtained by dissolving solid commercial chloroïodide of zinc, a moist

white salt, in somewhat less than its own weight of water, and then adding sufficient metallic iodine to give the solution a deep sherry-brown color. I prefer Grübler's preparation of the chloroïodide.] These solutions remain for a long time unchanged, especially when kept in the dark. The reaction succeeds best when the sections are placed directly in the concentrated reagent.

5. Recently a number of reagents containing iodine have been recommended by Mangin (VII), which act in the same manner as chloroïodide of zinc and seem to be, in part, more delicate than it.

Of these reagents I have used with good results a *calcium-chloride-iodine* solution, and, instead of following Mangin's somewhat more elaborate method, have prepared it by adding about .5 gram of potassium iodide and .1 gram of iodine to 10 ccm. of a concentrated solution of calcium chloride and then, after gentle warming, separating the solution from the excess of iodine by filtering through glass-wool.

This solution, in which the sections should be placed directly, colors lignified membranes yellow to yellow-brown, but pure cellulose walls become first rose-red and, after a time, violet. According to Mangin, it should be kept in the dark.

By the aid of *iodine-phosphoric acid* recommended by Mangin, one obtains a very deep violet coloring of cellulose walls, while lignified and suberized walls are colored yellow or brown. This reagent is prepared by adding a small quantity of potassium iodide (about .5 gram to 25 ccm.) and a few crystals of iodine to a concentrated aqueous solution of phosphoric acid, and gently warming the whole. The sections should be freed of all water adhering to their surfaces, by means of filter-paper, before being placed in this solution.

Mangin also recommends mixtures of aluminium chloride or stannic chloride with iodine and potassium iodide. For the manner of preparing and using these solutions, Mangin's work may be consulted.

[Mangin states (VIII) that the most important and most characteristic reaction of cellulose is its conversion into *hydrocellulose* or amyloid. This conversion is not certainly accomplished by acids, but the best results are obtained by treating the cellulose with a saturated alcoholic solution of *sodium* or *potassium hydroxide* and then transferring it to absolute alcohol. Cuprammonia also produces the same result. The above-described reagents for cellulose act promptly with hydrocellulose.]

247. Their behavior with *staining media* can also be used for the recognition of pure cellulose membranes. These also serve in delicate sections, microtome sections or the like, to bring out better the network of cell-walls.

Hæmatoxylin is especially adapted to this purpose, Giltay (I) having first observed the fact that it stains deeply only the unlignified and unsuberized membranes. It may be used in very various solutions (e.g. in the so-called Bohmer's (cf. § 315) or in Delafield's (cf. § 314) solution). These stain pure cellulose walls deep violet, while lignified and suberized membranes remain at first uncolored, or are stained yellow or brown. In most sections an exposure of a few minutes is sufficient for a deep staining of the membranes.

248. The writer has used hæmatoxylin with the best results for the recognition of the closing membrane of bordered pits (Zimmermann IV). With the wood of *Coniferæ* it is sufficient to leave the sections for fifteen minutes in Böhmer's hæmatoxylin solution (cf. § 315) to obtain a deep staining of the "tori" of the bordered pits, which, naturally, come out most sharply after clearing in Canada balsam (cf. §§ 14-22).

249. *Aniline blue* and *methyl blue* may also be used for staining cellulose walls. These produce an intense stain in an hour with microtome sections, which is not affected by alcohol, clove-oil, or xylol, so that the preparations may be mounted in Canada balsam. A solution of *Berlin blue* acts in the same way. This is prepared by allowing 1 gram of soluble Berlin blue and .25 gram of oxalic acid to stand several hours with a little distilled water, then adding 100 ccm.

of water and filtering (cf. Strasburger I, 622). To obtain a sufficiently deep stain, the solution must usually be allowed to act for several hours. It is not washed out by alcohol. [According to Mangin (VIII), pure cellulose is readily stained by many of the azo-colors, as by orseillin BB, croceïn and naphtol black in an acid solution, or by Congo-red and benzo-purpurin in an alkaline solution. Several of the dyes recommended heretofore for cellulose walls really stain only the pectic constituents of cell-membranes (cf. § 292). Such are methylene blue, aniline brown, and chinolin blue.]

250. A very deep and permanent staining of the wall is obtained, according to Van Tieghem and Douliot (I), by placing sections, after all cell-contents have been removed by *eau de Javelle* and caustic potash, and after thorough washing, first in a dilute solution of *tannin* for one to two minutes and then, as quickly as possible, in a very dilute solution of *ferric chloride*. They are at once removed from the latter solution and enclosed in glycerine or Canada balsam. All the membranes are then stained a deep black.

For staining the younger membranes of microtome sections, I have lately found *Congo-red* well adapted. I allow it to act in concentrated aqueous solution, for 24 hours, upon the sections, and then wash them in alcohol and mount in Canada balsam.

2. Lignified Membranes.

251. Lignified membranes are distinguished from those of pure cellulose by being insoluble in cuprammonia and by being colored yellow or brown by *iodine and sulphuric acid* or *chloroïodide of zinc*. It was formerly generally believed that this difference in chemical relations of lignified walls is due to the incrustation of the cellulose with a substance richer in carbon, *lignin*. And in fact lignified membranes give the reactions of pure cellulose after treatment with Schulze's macerating mixture (cf. § 9). According to Mangin (VII), the same thing occurs after treatment with *eau de Javelle*.

252. Recently the attempt has been made in several quarters to reach more accurate conclusions concerning the chemical composition of lignified membranes, and especially as to the constitution of lignin.

Wholly trustworthy results can, of course, only be reached by macrochemical investigations with exact quantitative analysis. In this connection should be mentioned the recent researches of Lange (I and II), who has isolated from the woods of the beech, oak, and fir, two compounds of an acid character, "*lignic acids*," which may, however, possibly come from a single substance. Lange also obtained various by-products concerning whose significance nothing is yet known.

253. There is also widely distributed in lignified membranes a gum-like substance which Thomsen has called *wood-gum*. It may be extracted with a 5% solution of caustic soda and then precipitated from this solution with 90% alcohol. On hydrolysis wood-gum yields either arabinose, $C_6H_{12}O_6$, or xylose, $C_5H_{10}O_5$.

Wood-gum and both of its derivatives above mentioned take a cherry-red color on warming with phloroglucin and hydrochloric acid. But Allen (I, 39) has shown that the phloroglucin reaction about to be described is not to be referred to the wood-gum ; for, on one hand, the reaction takes place with lignified membranes in the cold, and, on the other, the colors which appear in the different reactions show very different spectroscopic relations.

254. Attempts have also been made to determine the chemical constitution of lignified membranes by microchemical studies. Especially Singer (I) and, more recently, Hegler (I) have tried to prove that *coniferin* and *vanillin* always occur in lignified walls. This view is based chiefly on a series of color-reactions which lignified walls give with various aromatic compounds. I give a compilation of the chief of these reactions with remarks on their application, which should receive notice here because the reactions may be used with good results for the microchemical recognition of lignification.

SPECIAL METHODS.

TABLE TO ACCOMPANY § 254.—COLOR-REACTIONS OF LIGNIFIED CELL-WALLS.

Reagent.	Chemical Formula.	Manner of Using	Color of Lignified Walls.	Literature.
Phenol	C_6H_5OH	To the concentrated aqueous solution is added as much $KClO_3$ as it will dissolve, and this solution is added simultaneously with HCl	blue or greenish blue	v. Höhnel, II; Singer, I; Molisch, III, 304
Thymol	$C_6H_3.CH_3.C_3H_7.OH$	A 20% alcoholic solution is diluted with water until thymol begins to be precipitated; then solid $KClO_3$ is added in excess and the whole is filtered after several hours. The solution is used simultaneously with HCl	blue or blue-green	Molisch, III, 303
Resorcin	$C_6H_4(OH)_2$	Alcoholic solution + HCl	blue-violet	Wiesner, III, 65; Ihl, II
Orcin	$C_6H_3.CH_3.(OH)_2$	Alcoholic solution + HCl	dark red	Ihl, II
Phloroglucin	$C_6H_3.(OH)_3$	Aqueous or alcoholic solution added simultaneously with HCl	red-violet	Wiesner, III
α-Naphtol	$C_{10}H_7OH$	15% alcoholic solution added simultaneously with concentrated HCl	blue-green	Molisch, III, 305
Aniline sulphate	$(C_6H_5NH_2)_2.H_2SO_4$	Concentrated aqueous solution added simultaneously with H_2SO_4	golden yellow	Wiesner, III
Aniline chloride	$C_6H_5NH_2.HCl$	Added in aqueous or alcoholic solution simultaneously with HCl	golden yellow	v. Höhnel, I, 527
Tholuidendiamine	$C_6H_3.CH_3.(NH_2)_2$	Concentrated aqueous solution + HCl	dark orange	Hegler, I, 33
Indol	$C_6H_4.(CH)_2.NH$	Aqueous solution + mixture of 1 vol. concentrated H_2SO_4 and 4 vol. water	cherry-red	Niggl, I
Skatol	$C_6H_4.C_2H_3.CH.NH$	Concentrated alcoholic solution diluted with some HCl	violet	Mattirolo, I
Thallin sulphate	$(C_9NH_{10}.OCH_3)_2.H_2SO_4$	Concentrated solution in mixture of 1 vol. alcohol + 1 vol. water. Sections first placed in alcohol. Solution should be freshly prepared	yellow to dark orange-yellow	Hegler, I, 33
Carbazol	$(C_6H_4)_2.NH$	First a few minutes in warmed concentrated alcoholic solution, then in a drop of HCl or of a mixture of 1 vol. H_2SO_4 + 1 vol. water	red-violet	Mattirolo, I; Nickel, 1 59

255. According to Hegler (I, 40), all of these compounds that have been tested give the same color-reactions with coniferin or vanillin or with a mixture of both substances; and this author therefore considers it demonstrated that both compounds occur constantly in lignified membranes, and that they are the cause of the above described color-reactions.

Of the above enumerated reagents thallin and phenol deserve especial attention, since, according to Hegler (I), the former gives the described color-reaction only with vanillin, the latter only with coniferin.* Since thallin colors vanillin yellow, but phenol colors coniferin blue, by the use of a mixture of the two reagents one may draw certain conclusions as to the relative abundance of the two substances, according as the color produced is yellower or bluer. Hegler (I, 58) has used for the same purpose a solution prepared by mixing .5 gram of thallin sulphate, 1.3 grams of thymol, 2 ccm. of water, 26.5 ccm. of alcohol, and .5 gram of potassium chlorate, and diluted for use with its own volume of hydrochloric acid of specific gravity 1.124. This author draws from his studies carried on with this reagent, the conclusion that the younger xylem-cells contain more coniferin than vanillin, but that the older ones are rich in vanillin and less so in coniferin.

256. On the other hand, it is to be noticed that, besides vanillin and coniferin, other substances containing the aldehyde group give with the compounds mentioned identical or similar color-reactions to those of lignified cell-walls; according to Ihl (I), cinnamic aldehyde, and, according to Nickel, salicylic aldehyde. It may therefore be considered as good as proven that the color-reactions of lignified membranes depend upon the presence of one or of various compounds belonging to the aldehyde group. Seliwanoff's observations also support this view. According to these, lignified walls are, on one hand, colored red by a solution of fuchsin decolorized with sulphurous acid, and, on the other hand, no longer give the reactions of lignified membranes with phloro-

---

* The correctness of Hegler's statement that thymol also gives no color with vanillin has lately been disputed by Molisch (I, 48, Note 3).

glucin etc. after treatment with hydroxylamine, which chemically unites with aldehydes, destroying the aldehyde group (cf. Nickel II, 755).

In what relation these aldehyde-like compounds stand to the so-called lignin cannot at present be stated.

257. For the microchemical recognition of lignification, besides the behavior with cuprammonia and with iodine solution already described (§ 251), the color-reactions given in the foregoing table may be used. Of these reagents, *aniline sulphate* and *phloroglucin* are especially good. The colors produced by these substances last but a short time, while *thallin* gives permanent colors and is therefore adapted to the making of permanent preparations, which may be mounted in glycerine-gelatine or in Canada balsam.

258. Various pigments may also render good service in the study of lignified walls. These behave quite differently from unlignified walls with staining media and therefore certain staining solutions may be well used for the distinction of the different sorts of membranes.

259. For the staining of lignified membranes *fuchsin* has shown itself especially useful, and has already been recommended by Van Tieghem and Berthold for this purpose. I obtained very beautiful permanent preparations, in which only the lignified walls were stained deep red, by leaving the microtome sections first for a quarter of an hour or longer in an aqueous solution of fuchsin, and then washing them for a short time in a solution of picric acid, such as Altmann's, which contains one part of a concentrated alcoholic solution of picric acid to two parts of water (cf. § 345). This makes them dark-colored, and much of the color is then washed out with alcohol; finally they are passed into xylol and xylol-Canada balsam.

If a *double staining* is desired, it may be obtained by placing the sections, after the washing in alcohol, for an hour in a suitable solution of hæmatoxylin, such as Böhmer's (cf. § 315), aniline blue, methyl blue, or Berlin blue. After washing again in alcohol and mounting in Canada balsam, one obtains preparations in which the lignified membranes are stained deep red and pure cellulose membranes violet or blue.

260. *Acid fuchsin** gave me similar preparations, which were washed either with running water or with Altmann's picric acid solution (cf. §§ 345-6). Aniline-water-*safranin* also gave me preparations in which only the lignified and suberized walls were stained after being washed out with acid alcohol (cf. § 268), after acting for at least an hour. *Gentian violet* acted in the same way when used according to Gram's method (§ 321). All these preparations may be well preserved in Canada balsam, and the dyes may be used for double staining in the way above described.

261. The staining methods named may be very well used for the demonstration of the course of the vascular bundles in whole parts of plants, such as leaves or thin stems. I obtained very instructive preparations with a concentrated aqueous solution of fuchsin under which I cut off the part to be stained, so that the vascular bundles were mostly deeply stained in a relatively short time. When this was accomplished, I placed pieces of the objects in alcohol until the chlorophyll was completely removed, cleared them in clove-oil, and transferred to Canada balsam. Especially favorable objects for study are found in the leaves of *Secale cereale* and of *Impatiens parviflora*. In the latter only the tracheal elements were colored red, and could be very clearly seen even in the thicker parts.

3. The Cuticle and Suberized Membranes.

262. Until recently it has been generally assumed that suberization is due to the incrustation of the cellulose wall with a fat-like substance commonly called suberin. This assumption has been based especially on the observation of Fr. von Höhnel that suberized membranes are colored red-violet with chloroïodide of zinc after treatment with an aqueous solution of caustic potash. But the recent

---

* This dye is also called " Fuchsin S," and by Dr. Grübler, " Fuchsin S after Weigert."

researches of Gilson (I) have made it at least very questionable if there is any cellulose present in suberized walls. This author isolated from the cork of various plants an acid, *phellonic acid*, which, as well as its potassium salt, becomes rose- or copper-red with chloroïodide of zinc. Gilson believes that the cause of the violet or rather reddish color which these membranes treated with caustic potash assume with chloroïodide of zinc is to be sought in the presence of potassium phellonate. In fact, the staining described does not appear if the walls are extracted with boiling alcohol after treatment with caustic potash, and before the addition of the chloroïodide. The color which appears after preliminary treatment with chromic acid is probably due, according to Gilson, to the formation of free phellonic acid. The absence of color after treatment with cuprammonia is due, not to the solution of cellulose, but to the conversion of potassium phellonate into the copper salt, which takes a yellowish-brown, very slightly characteristic color with chloroïodide of zinc. Finally, the presence of cellulose in the suberized wall is rendered improbable by the fact that the whole suberin lamella may be made to disappear, according to Gilson, by long continued treatment with a 3% boiling alcoholic solution of potassium hydrate, which does not recognizably attack cellulose.

263. Besides *phellonic acid*, already described, to which Gilson assigns the formula $C_{22}H_{44}O_2$, two other acids have been isolated from the cork of *Quercus Suber* by the same author, *suberic acid* ($C_{17}H_{30}O_3$) and *phloionic acid* ($C_{11}H_{21}O_4$?). It still remains undetermined in what form these acids are contained in suberized membranes. But it is not probable that they occur as true glycerine ethers, since the suberin lamella is insoluble in all solvents for fats and could not be melted by Gilson when heated up to 290° C. The view suggested by Kügeler (I, 44) that suberin is so difficult of solution because the suberin molecules are enclosed between cellulose molecules is untenable, since it is shown that the suberin lamella contains, at most, only traces of cellulose. Therefore, at present, Gilson's view that suberin consists

of compound ethers or of condensation- or polymerization-products of various acids seems to have most in its favor.

[Van Wisselingh has lately (1) found that most of the constituents of suberin melt at a temperature below 100° C., but are deposited in a substance that does not melt and must first be removed. He finds that the suberin constituents are mostly soluble in *chloroform*, and believes that they consist of various fatty substances with glyceril or other compound ethers.]

264. In this connection the optical relations of suberized walls are worthy of attention, since they enable us to draw some conclusions as to the form in which the substances in question occur in the membranes. The suberized membranes, as well as the cuticle, show a pretty strong double refraction, and their optical axes are usually placed in the reverse position to those of the pure cellulose wall. But this double refraction disappears completely, as Ambronn has shown (1), on heating to 100° C., reappearing as before on cooling. This may be easily seen by heating cross-sections of the leaf of *Agave americana* in glycerine until the fluid boils and then examining them with a polarizing microscope.

The optical relations of cork clearly compel the view that its double refraction is due to the presence of regularly arranged particles of crystalline form, which melt on heating and, on subsequent cooling, recrystallize in the same regular arrangement.

265. It remains to be determined by further studies whether all suberized membranes have the same composition, and to what extent the external layers of the epidermal cells, the cuticle and the cuticular layers, agree in material constitution with the suberin lamella of cork cells.

266. As has been long known, suberized membranes, as well as cuticularized ones, show the following relations to chemical reagents:

They are insoluble in cuprammonia, are never colored blue or violet by iodine and sulphuric acid or by chloro-

Iodide of zinc, but always yellow or brown; and are insoluble in concentrated sulphuric acid.

But, according to Fr. von Höhnel's researches (I), the following reactions are especially characteristic :

Concentrated *caustic potash* solution causes in the cold a yellow coloring of suberized membranes, which increases in intensity when they are warmed in this fluid. The suberized membranes take, at the same time, a lineate or granular structure which becomes plainer on further warming. On boiling in the same solution, the large yellow drops which are formed often escape entirely from the membrane.

Schulze's macerating mixture (cf. § 9, 1) is resisted longest by suberized walls, of all the modifications of cellulose; but they finally run together, on long boiling in the fluid, into oil-like drops whose substance is termed *ceric acid* and is soluble in hot alcohol, ether, chloroform, benzol, and a dilute solution of caustic potash, but insoluble in carbon bisulphide.

Concentrated chromic acid either does not dissolve the suberized membranes at all, or only after acting for a day; while all the other modifications of cellulose, except funguscellulose, are dissolved by this acid in a short time.

267. For distinguishing between suberized and lignified cell-membranes, chlorophyll and alcannin may be used.

Correns (II, 658, note) first recognized the fact that *chlorophyll* stains the cuticle and suberized membranes deep green, while the lignified and pure cellulose walls are not colored. For this purpose, a freshly prepared alcoholic solution of chlorophyll, as concentrated as possible, should be allowed to act on the sections in darkness for a quarter of an hour or longer. The sections may then be examined in water. They cannot be preserved by ordinary methods.

For staining with *alcannin*, a solution of this substance in 50% alcohol is used, in which the sections are left for several hours or longer. All suberized membranes and the cuticle take a red color which is not so deep as with any fats which may be present, but is always clearly visible.

Both of these stains are of interest as indicating anew that fat-like substances are deposited in the cuticle and cork.

267a. I have recently found that, besides alcannin, which gives a very deep staining of the cuticle after long action, *osmic acid* and *cyanin* may be also used for the recognition of suberized membranes. For this purpose, I dissolve cyanin in 50% alcohol and add an equal volume of glycerine. Preliminary treatment of sections with *eau de Javelle* is generally to be recommended, as it destroys the tannins which hinder the staining. It also causes the lignified walls to lose their power of staining, while suberized ones are as deeply stained as in the fresh condition, even after being exposed for a day to its action (cf. Zimmermann VII).

268. In their behavior with staining media, the cuticularized and suberized membranes show in many respects an agreement with lignified ones. This is especially true of the so-called suberin lamella of cork-cells; but the true cuticle is often less easily stainable. But it is usually easy to stain the cuticular layers differentially, especially in thick-walled epidermal cells; and double stainings, in which these layers are differently colored from the cellulose layers lying beneath, may be obtained. But it must be remarked that these stains do not always act with the same precision, in all cases, as with lignified walls; and different plants do not appear to behave in the same way in this respect. I recommend as suitable objects for study, the leaves of *Clivia nobilis* or *Agave americana*. On these the following stainings and double stainings may be readily carried out. The statements as to time refer to microtome sections.

a. Safranin.

269. The best staining medium for suberized walls is aniline-water-safranin, prepared by mixing equal volumes of aniline-water and a concentrated alcoholic solution of safranin. I allow this to act for half an hour or longer on the sections, then cover them with acid alcohol,* which is quickly replaced

---

* That is, alcohol to which is added about .5% of the ordinary chemically pure HCl.

by alcohol. They are washed with the latter until no more color is given off, and then transferred to Canada balsam in the usual way. Especially if the washing with acid alcohol is just right, only the lignified and suberized cell-walls are stained in these preparations, in which the former show a bluish, the latter a rather yellowish, color.

If it is desired to stain the cellulose walls also, this may be done by one of the following methods:

α. Methyl Blue.

The sections, stained with safranin and washed with alcohol, are placed in a concentrated aqueous solution of methyl blue, in which they remain a quarter of an hour or longer. They are then washed in alcohol and mounted in Canada balsam. The cellulose walls are then stained blue, the suberized and lignified ones, red.

β. Aniline Blue

This must be used in aqueous solution which must be first washed off with water after the staining, since turbidity readily results from the direct addition of alcohol. Otherwise it is used like methyl blue.

γ. Hæmatoxylin.

Böhmer's hæmatoxylin (§ 315) may well be used for double staining with safranin. This is allowed to act for a few minutes on sections stained with safranin and washed, is then washed off with water, and the sections are mounted in Canada balsam. Sections thus treated show the lignified and suberized walls red, and the cellulose walls violet.

b. Gentian Violet and Eosin.

270. In the so-called Gram's staining process (§ 321) with gentian violet, only the lignified and suberized walls remain stained after thorough washing with clove-oil; but a fine double staining may be obtained by proceeding according to Gram's method and adding to the clove-oil used in washing

a little eosin, which dissolves readily in it. The eosin at once stains the cellulose walls a beautiful red, while not changing the staining of the other walls.

c. Ammonia-fuchsin.

271. As was first recognized by Van Tieghem, ammonia-fuchsin is well adapted to staining suberized and lignified membranes. It is prepared by adding ammonia to a not too concentrated alcoholic solution of fuchsin, until the solution becomes straw-yellow after a little shaking. The solution should be filtered after a few days, but can be kept only a few weeks, even in well closed bottles.

A double staining may be had by placing the sections first in the above described ammonia-fuchsin solution for a few minutes, and then passing them directly to an aqueous solution of methyl blue, in which they are left a quarter of an hour or longer, then washing with alcohol and mounting in Canada balsam.

d. Cyanin and Eosin.

272. If sections are placed for several hours in a freshly prepared, very dilute aqueous solution of cyanin, which may be prepared by adding 20 drops of a concentrated alcoholic solution to 100 ccm. of water, the lignified and suberized membranes appear beautifully blue after washing in alcohol. If clove-oil containing eosin be used in transferring to Canada balsam, a fine double staining is obtained. The modified walls are blue, the cellulose walls red.

I obtained also a deep staining of the cuticle by leaving sections for a considerable time in a solution of cyanin in 50% alcohol and then washing out the stain with glycerine.

4. Gelatinized Cell-walls, Plant-mucilages, and Gums.

273. The so-called gelatinized membranes are distinguished from cellulose walls chiefly by their different physical character, their strong power of swelling; and indeed there occur all stages between pure cellulose, which takes up little water, and the gums which are wholly soluble in water, like gum arabic. Part of these substances are formed from cellulose, but most of them are formed by the plant directly as mucilages. It should be observed that the occurrence of vegetable mucilages and gums within the plant is not at all restricted to the cell-wall, but they may also be formed within the protoplasm. But it has seemed to me best to discuss all these bodies together here, in view of their undoubted relationships, which may lead one, with Beilstein (I, 877), to group them under the designation *gums*.

274. So far as the chemical relations of the gums are concerned, it should first be observed that most of them, so far as they have been analyzed, agree in their percentage composition with cellulose and thus correspond to the formula $C_6H_{10}O_5$. But, on the other hand, they differ considerably from cellulose in their chemical relations, and also show great differences among themselves.

Thus some of them are colored blue by *iodine* alone, others only by iodine and sulphuric acid or *chloroïodide of zinc*, and still others are colored only yellow or not at all by iodine preparations.

A part of the gums are soluble, a part quite insoluble, in *cuprammonia*.

On oxidation with nitric acid, a part of them give oxalic acid, $(COOH)_2$, a part, mucic acid, $(CHOH)_4.(COOH)_2$, a part, both acids.

Unfortunately the chemical characters of the various gums are not determined with sufficient exactness to make possible a strictly scientific grouping of them. But in the following account some remarks on the general methods of recognizing the gums may be in place, and then the chief chemical characters, and especially the microchemically applicable reactions, of the gums which have been studied in detail may be brought together.

275. For the microchemical recognition of the gums their strong power of *swelling* in water may first be used. To follow the process of swelling exactly with the microscope, one may first place the objects in absolute alcohol, in which all the gums are insoluble and do not swell, and then

gradually allow water to enter from the edge of the coverglass.

The dissimilar behavior of the gums with iodine solutions and with cuprammonia has already been mentioned. Besides these, *corallin* may be used in many cases in the study of gelatinized cell-walls and gums, since many of them are deeply stained by it. Since it is practically insoluble in water, it may be dissolved in a concentrated solution of soda. This solution gradually decomposes, but preserves its staining power for a long time (cf. also § 289).

Characteristic stainings of plant-mucilages are often obtained with Hanstein's aniline mixture.*

a. Amyloid.

276. The substance known by the name amyloid occurs in the seeds of various plants (*Tropæolum majus, Impatiens Balsamina, Pæonia officinalis,* many *Primulaceæ,* and others) and constitutes a reserve material which goes into solution on the germination of the seed.

Amyloid is characterized by being colored blue by *iodine* solutions, the best adapted for this reaction being, according to Nadelmann (I, 616), a dilute solution of iodine and potassium iodide, since a concentrated solution of the same substances colors it brownish orange, and fresh tincture of iodine does not generally color it at all at first.

In cuprammonia amyloid is insoluble.

Its behavior with *nitric acid* is also characteristic.

In an acid which contains 30% of HNO_3 (spec. gravity 1.285), amyloid at once swells strongly, and after a time becomes entirely dissolved (cf. Reiss I, 735, 737, 739).

The amyloid contained in the seeds named is not identical with the compound prepared from cellulose by treatment with acids (§ 246), which has often been termed amyloid (cf. Beilstein I, 863, 882). Amyloid is distinguished from reserve-cellulose (cf. § 286) by the reactions already de-

---

* [This consists of an alcoholic solution of equal parts of fuchsin and methyl violet.]

scribed and by the fact that its hydrolytic splitting with sulphuric acid yields no seminose, but most probably glucose (cf. Reiss I, 761).

[Winterstein's (I) recent researches give results which differ in several respects from those of Reiss. He finds amyloid soluble in *cuprammonia* after a day. From this solution it is not precipitated by acids, but is thrown down by alcohol. Its composition seems to correspond to the formula $C_{12}H_{30}O_{16}$, and it appears to belong to Tollens' group of Saccharo-colloids, though it is not certain that it is a single compound. In spite of its bluing with iodine, it cannot be regarded as very nearly related to starch.]

b. Wound-gum.

277. The name wound-gum is commonly given to a substance which, according to Temme's researches (I), is very abundantly secreted in the vessels by the surrounding starch-cells, in natural and artificial wounds, and, like tyloses, closes their cavities. This wound-gum agrees, according to Temme, with many sorts of gums in that it yields oxalic and mucic acid on oxidation with nitric acid. But it differs essentially from all gums in not swelling in water and in being insoluble even in caustic potash and sulphuric acid. As has been recognized by Temme, wound-gum is stained deep red by *phloroglucin* and hydrochloric acid. Molisch showed later (IV, 290) that it behaves just like lignified membranes with *aniline sulphate*, *metadiamidobenzol*, *orcin*, and *thymol*; and he believes that wound-gum contains vanillin in solution (cf. § 254).

c. The Gelatinous Sheaths of the Conjugatæ

278. In many *Zygnemaceæ* the whole surface of the cell-filaments is surrounded by a colorless covering, a "gelatinous sheath," while in the *Desmidiaceæ* the excretion of jelly is often limited to distinct regions on the membrane (cf. Klebs II, and Hauptfleisch I).

Since the refractive index of these jelly-sheaths differs but little from that of water, they can be well recognized, when unstained, only by the aid of strong objectives. But with lower powers they stand out clearly when the algæ are placed in very finely rubbed India ink, according to the method proposed by Errera (III). For this purpose, enough of the genuine Chinese "India ink" may be rubbed up directly on the slide to give the drop a dark-gray appearance, and then the alga to be studied is placed in it.

No trustworthy statements can yet be made as to the chemical composition of these jelly-masses; and it need only be said that they give no cellulose reactions either with iodine and sulphuric acid or with chloroïodide of zinc, and that they are always sharply defined against the cellulose wall and are not in genetic connection with it.

A number of observations, made especially by Klebs (II) on the gelatinous sheaths of the *Zygnemaceæ*, deserve more detailed notice, as they show that these must possess a very complicated organization.

279. Klebs first established the fact that the gelatinous sheaths always consist of two different substances, one of which can be extracted with hot water and is pretty deeply stained by certain dyes, like methylene blue, methyl violet, and vesuvin; while the substance which is insoluble in hot water remains quite colorless with these stains. After staining with one of the colors above named, delicate rods are seen in the sheaths, which often appear united into a fine network at the ends which are directed toward the cell-lumen (cf. Fig. 33, *I*). The same structure can also be made visible by other means, especially by alcohol. It is evidently due to the fact that the different substances are unequally distributed in the jelly-sheath.

280. A further remarkable character of the jelly-sheaths consists in the fact that, after the deposition in them of certain precipitates, for instance, of Berlin blue, these are thrown out, together with a greater or less part of the water-soluble substances of the sheath, with swelling of the latter (cf. Fig. 33, *II* and *III*). This "throwing off of the gelatinous sheath" begins with an accumulation of the previously evenly scattered particles into evident granules (cf.

Fig. 33, *III*), which are held together by the mucilage separated with them and finally thrown off with them.

This expulsion may be caused by various precipitates. A very suitable one is chrome yellow ($PbCrO_4$), which is precipitated in the membranes by placing the algæ, held together by a thread, in a .25% solution of potassium chromate (K_2CrO_4), then rinsing quickly in water, and finally transferring them to a .25% solution of lead acetate. To obtain a heavy precipitate, this proceeding may be several

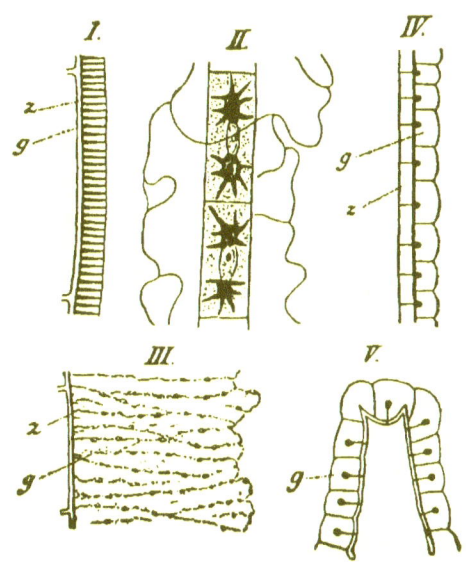

FIG. 33.—*I*, membrane and gelatinous sheath of *Zygnema* sp. (× 580). *II*, two *Zygnema*-cells after deposit of chrome yellow (× 245). *III*, membrane and gelatinous sheath of *Zygnema* after deposit of chrome yellow (× 245). *IV*, the same of *Pleurotænium Trabecula*, after staining with fuchsin (× 950). *V*, the same of *Staurastrum bicorne*, after staining with gentian violet (× 950). *z*, cell-membrane; *g*, gelatinous sheath. *I* to *III* after Klebs; *IV* and *V* after Hauptfleisch.

times repeated. But the expulsion takes place the more rapidly the less chrome yellow is deposited, and may not be completed for several days if the deposit be large.

Finally, it may be remarked that this expulsion is not directly dependent on the life of the protoplasm, and may occur in dead individuals, under some circumstances.

281. Klebs has also established the remarkable fact that the gelatinous sheaths increase markedly in density in a solution of glucose and peptone by the deposit of a substance whose composition is not yet known.

This "thickening" of the gelatinous sheaths occurs, however, only when soluble albuminoids and a sugar are simultaneously present in the surrounding fluid, and, like the expulsion described, is independent of the life of the protoplasm.

282. According to the investigations of Hauptfleisch (I), the gelatinous formations of the *Desmidiaceæ* consist, on the other hand, of single prisms or caps, each of which covers a pore in the cell-wall. These pores are occupied by threads of protoplasm which commonly terminate externally in globular swellings which penetrate to a greater or less distance into the gelatinous covering, in different species (cf. Fig. 33, *IV* and *V*).

For the observation of these structural relations, this author recommends that at first dilute, and then gradually more concentrated solutions of safranin, fuchsin, gentian violet, methylene blue, or methyl violet, be allowed to run from the edge to the living algæ under a cover-glass, and that the changes in the jelly during the action of the stain be followed. Then the changes may be followed backward by careful washing of the specimens.

The presence of two different substances in the gelatinous covering has been disputed by Hauptfleisch for the Desmids.

5. Fungus-cellulose.

283. The membranes of the fungi show very varying relations. In a number of species they give the normal cellulose reactions, and this is especially the case in young stages (cf. de Bary II, 9). But in most fungi they differ from pure cellulose membranes in being insoluble in cuprammonia and in being colored only yellow or brown by iodine and sulphuric acid or by chloroïodide of zinc. They also show great powers of resistance to alkalies and acids in general. But since, on the other hand, they do not show the reactions for lignification or suberization, we are compelled at present to regard them as a special modification of cellulose, which is commonly termed fungus-cellulose.

It should be remarked that, according to the researches of K. Richter (I), the membranes of a large number of fungi give the reactions for pure cellulose after being first treated for a long time with caustic potash. But in many cases the caustic potash must act for a week.

On the other hand, W. Hoffmeister (I, 254) has lately obtained from the fructification of *Boletus edulis*, by the use of methods always successful with the higher plants, no compounds giving the reactions of cellulose. The membranes of this fungus are, according to his researches, characterized by being completely soluble in concentrated hydrochloric acid and caustic potash.

284. *The Membranes of the Bacteria.*—There can now be no doubt that the Bacteria possess a solid membrane. In most cases its presence may be readily demonstrated by plasmolyzing the organisms (cf. §§ 431 and 463).

No reliable statements can be made at present as to the chemical constitution of these membranes. They seem, moreover, to consist in part of cellulose; at least, W. Hoffmeister (I, 253) has isolated a substance reacting like cellulose from a species of *Bacillus* not exactly determined.

6. Paragalactan-like Substances (Hemicelluloses).

285. Reiss and E. Schulze have shown that, especially in the cell-walls of seeds with considerable thickenings of the walls, carbohydrates occur which differ essentially from cellulose and are dissolved at germination, like the other reserve materials of the seed. One of these substances is called by Reiss reserve-cellulose, another, by Schulze, paragalactan. But it is probable that various related compounds exist. All these substances can at present best be grouped under the name proposed by E. Schulze, " Paragalactan-like compounds."

[Schulze's later studies (II) afford ground for distinguishing this group of substances from cellulose as *hemicelluloses*. He finds that they become soluble, with the formation of glucose, through the action of hot dilute *mineral acids*, by which true celluloses are not affected. They are dissolved by dilute alkalies and by *cuprammonia* after brief treatment with hot dilute hydrochloric acid, too short to cause their solution.]

a. Reiss' Reserve-cellulose.

286. The so-called reserve-cellulose has been prepared by Reiss (I) from the endosperm of *Phœnix dactylifera*, *Phytelephas*, and various other seeds with strongly thickened cell-walls. It differs from the ordinary cellulose especially in the products resulting from hydrolysis with sulphuric acid. There is first formed a compound corresponding to dextrine, but lævo-rotary (*seminin*), and then a dextro-rotary sugar which reduces Fehling's solution and is fermentable (*seminose*) and especially characterized by the fact that it forms with phenylhydrazin acetate ($C_6H_5NH.NH_2$) an hydrazon, which may be obtained in crystalline form, of the composition $C_{18}H_{22}N_2O_5$, probably according to the reaction: $C_6H_{12}O_6 + C_6H_5N_2 = C_{12}H_{18}O_5N_2 + H_2O$. Reserve-cellulose cannot be distinguished microchemically from ordinary cellulose and behaves quite like pure cellulose with *iodine* solutions and *cuprammonia*.

An exception is shown only by the cell-walls of the endosperm of *Paris quadrifolia* and *Fœniculum officinale*, which are insoluble in cuprammonia, although they give seminose on hydrolysis and must therefore be regarded as reserve cellulose.

[Schulze finds (II) that this substance shows the characters of other hemicelluloses (cf. § 285) and should be placed among them, with the name *mannose*.]

b. Paragalactan.

287. The name paragalactan has been given by E. Schulze (cf. Schulze I, and Schulze, Steiger, and Maxwell, I) to a compound recognized in the thickenings of the walls of the cells of the cotyledons of *Lupinus luteus*, with true cellulose, and which very probably occurs in other *Leguminosæ*. It yields, on oxidation with nitric acid, mucic acid; on heating with dilute sulphuric acid, galactose ($C_6H_{12}O_6$) and a pentaglucose. It is also characterized by giving a cherry-red fluid on heating with phloroglucin and hydrochloric acid, while no color is produced in the cold. On heating, paragalactan is transformed by 1% hydrochloric acid or 1% sulphuric acid into sugar, while cellulose is attacked only by pretty concentrated solutions.

It is an important fact for the microchemical recognition of paragalactan that it is insoluble in cuprammonia and prevents the solution of the cellulose which occurs in the same membranes, while the latter is readily dissolved by cuprammonia after the removal of the paragalactan by boiling 2.5% hydrochloric acid. Paragalactan does not seem to be colored by chloroïodide of zinc; at least, membranes treated with this reagent showed only a slight bluing, while the remains of the membrane are deeply colored after the solution of the paragalactan.

[This substance also shows the characteristics of the hemicelluloses (cf. § 285). The pentaglucose which it yields besides galactose is probably arabinose, and it may therefore be called paragalacto-araban. It is very possible that it is a mixture of two substances, galactan and araban.]

c. Arabanoxylan.

[287a. Schulze finds (II) a hemicellulose in wheat and rye bran which yields, on hydrolysis, an arabinose and a xylose, and may therefore receive the above name.]

7. Callose, the Callus of the Sieve-tubes.

288. Until recently the name callus was generally given to a pretty strongly refractive substance which causes a more or less complete closing of the sieve-pores in old sieve-tubes, and finally covers the whole sieve-plate with a thick mass. Mangin has lately recognized (I–III) the more general distribution of this substance, especially in the membranes of various pollen-grains and pollen-tubes and in many fungi. It is, for instance, widely distributed in the

mycelium of the *Peronosporaceæ*, where it partly incrusts the cellulose walls and partly occurs in more or less pure condition in the interiors of the hyphæ and of the haustoria (cf. Mangin III). [The same author has also shown (IX) the presence of this substance in various cell-walls of Phanerogams which are incrusted with carbonate of lime, especially those of the cystoliths of the *Urticales* and of the calcareous hairs and pericarps of several *Borraginaceæ*. In the achenes of *Lithospermum, Cynoglossum*, etc., where it occurs without a deposit of lime, its occurrence seems to be related to the disappearance of the cell-contents and the gradual destruction of the parenchyma. He has also observed it in the walls of cells bordering tissues which have become suberized in consequence of injuries.]

This author calls this substance *callose*, a term which deserves preference, since the word "callus" is used, as is well known, in quite another sense.

288a. Callose gives the following reactions, according to Mangin (II): It is insoluble in water, alcohol, and cuprammonia, in the latter even after previous treatment with acids. But it is readily soluble in a cold 1% solution of *caustic soda* or potash, and is also soluble in the cold in concentrated *sulphuric acid*, as well as in concentrated solutions of calcium chloride and stannic chloride. Cold solutions of alkaline carbonates and of ammonia make it swell and give it a gelatinous consistency, but without dissolving it.

Callose also differs from cellulose in its behavior with various coloring matters. Mangin gives (V) a number of azo-colors which deeply stain cellulose in a neutral or feebly acid solution, but leave callose uncolored; they are especially *orseillin BB, azorubin, naphtol black*, and the *croceins*. On the other hand, callose is distinguished by its strong staining capacity with *corallin* and *aniline blue* and certain dyes belonging to the benzidines and tolidines.

289. *Corallin* or rosolic acid is best dissolved in a 4% or concentrated aqueous solution of soda (Na_2CO_3). The sections are placed for a short time in this solution and then examined in glycerine, when, if the staining has taken place properly, the deep-red pads of callose stand out sharply in the sieve-tubes. I have found it very useful to first over-stain the sections with corallin solution and then to wash them out with 4% soda solution, which quickly decolorizes all parts except the callose. This method has given especially good results with the fungi. Preparations stained with corallin cannot be long preserved.

290. *Aniline blue* has been recommended by Russow (I) for staining sieve-tube callose. It may be used in a dilute aqueous solution, which is allowed to act half an hour or longer on the sections. Overstained sections may be washed out with glycerine. They are properly stained when only the callose masses appear deeply colored. The "Schlauch-köpfe" of young sieve-tubes (§ 455) are also pretty deeply colored by aniline blue. To distinguish these from the callose, the preparations may be subsequently stained with an aqueous solution of eosin, in which they are left for a few minutes. After a brief washing in glycerine the entire contents of the sieve-cells, including the protoplasmic threads which penetrate the sieve-plates, are colored violet or red, while the callose pads remain deep blue. These preparations, as well as those with aniline blue alone, can be well preserved in glycerine-gelatine; or they may be transferred to Canada balsam in the usual way.

Since it often stains the protoplasm pretty deeply, aniline blue has usually given me much less instructive preparations than rosolic acid with fungi.

[**290a.** Mangin recommends (IX), for staining the callose of calcified membranes, a mixture of *soluble blue* extra 6B and *vesuvin*, or of the same blue and orseillin BB. These mixtures stain callose blue in a short time, the protoplasm and lignified elements being brown or violet, according to the mixture used. Where incrustations are not numerous, as on many leaves, large pieces of tissue may be freed from air by boiling alcohol, then placed in cold nitric acid until frothing ceases, then in cold water, in boiling alcohol, and finally in cold ammonia, to remove xanthoprotein and its derivatives. When the tissue is transparent enough, the ammonia may be neutralized with acetic acid, and the tissue placed in the staining fluid.]

291. The behavior of callose with *iodine* reagents, which has been exactly determined only for the callose pads of sieve-tubes, is also characteristic, and, according to Lecomte (I, 268), best brings out their intimate structure. Chloro-iodide of zinc stains callose brick-red or red-brown according to the proportion of iodine it contains, calcium chloride and iodine solution (cf. § 246, 5) stains it rose-red, or wine-red after previous staining with aniline blue, while the sieve-plates are colored violet.

8. Pectic Substances.

292. Mangin (IV–VI) has lately shown microchemically that pectic substances (pectin, pectose, pectic acids) are very widely distributed in the cell-walls of the most different plants, and that they form especially the so-called intercellular substance of unlignified and unsuberized membranes.

293. For the microchemical recognition of pectic substances Mangin uses (IV and VI) chiefly various coloring matters, *phenosafranin*, *methylene blue*, *Bismarck brown*, *fuchsin*, Victoria blue, violet de Paris (= methyl violet B), and rosolan (= mauveïn), and others. These do not color pure cellulose, but do stain pectic substances, as well in neutral solution as after slight acidification with acetic acid. But lignified and suberized membranes are also stained by these dyes. However, there remains a distinction between them and pectic substances in that the latter are quickly decolorized by alcohol, glycerine, and acids, while the former retain their color in these fluids. Mangin also gives a number of dyes which leave pectic substances uncolored in a neutral solution, while they deeply color lignified and suberized walls. Such colors are: acid green, acid brown, nigrosin, indulin, the croceins, and the ponceaux. Mangin obtained instructive double stainings by mixing one of these dyes with one of the previous group.

On the other hand, Mangin (V) has lately named a number of dyes which leave pectic substances uncolored, but stain cellulose or both cellulose and callose. To the former belong orseille red A, naphtol black, and the croceins; while Congo-red, azo-blue, and benzopurpurin stain cellulose and callose.

294. In order to show that the first described stainings really depend upon the presence of pectic substances, Mangin treated thin sections for 24 hours with cuprammonia and then washed them in water and in 2% acetic acid. On this treatment the cellulose is removed from the membranes and fills the intercellular spaces and the cell-cavities as a gelatinous mass. In consequence of it the membranes are colored not at all or but slightly yellow on the addition of chloroïodide of zinc, while a deep blue color appears in the interiors of the cells. The membranes, which now consist of pure pectic acid, are, however, deeply stained by safranin or methylene blue. It is sufficient to add a few drops of a solution of *ammonium oxalate* to cause the solution of the pectic acid membranes.

295. In order to show that the middle lamella of the so-called cellulose membranes consists of pectic acid or an insoluble salt of it, Mangin (VI) lets a mixture of one part hydrochloric acid and 4 to 5 parts alcohol act for 24 hours on thin sections, then washes them with water, and treats them with a weak (about 10%) solution of ammonia. After this has acted a short time, the sections may be separated into their constituent cells by gentle pressure. Mangin explains this by the supposition that the pectic acid is set free from its originally insoluble compounds by the action of the acid-alcohol, and is then dissolved by the ammonia solution. In fact, a gelatinous mass is precipitated from the ammoniacal solution on the addition of acid, which has all the characters of pectic acid. On the other hand, sections which were placed in lime- or baryta-water after the action of the acid-alcohol, showed no separation into their cells on subsequent treatment with ammonia, because the pectic acid had recombined into an insoluble salt with the alkaline earth.

296. Mangin (VI) obtained a deep staining of the middle lamella on placing thin sections of adult plant-organs in phenosafranin or methylene blue after treatment with the above mentioned acid-alcohol mixture. The middle lamella of pectic acid stains much more deeply than the pectic compounds mixed with cellulose of the thickenings of the cell-wall.

9. Ash- and Silica-skeletons of the Cell-wall.

297. The inorganic salts which incrust all vegetable cell-walls are in many cases present in such quantity that, after the destruction of all organic substances by burning, they still preserve the form of the original membranes.

Such ash-skeletons may easily be obtained by burning cross-sections of the stem of *Cucurbita Pepo* on the cover-glass. But they must be examined in the air, as they are at least partly soluble in water. These ash-skeletons consist chiefly of potassium and calcium salts. In other cases silicic acid also occurs deposited in great quantity in the membranes. For methods of recognizing this, see §§ 78–81.

10. On the Developmental History of the Cell-wall.

297a. In the study of the growth of cell-walls it is often important to stain the membranes without affecting the vitality of the cells. If the objects thus treated are then allowed to develop further in pure water, it would seem possible to distinguish the newly formed membranes or parts of membranes from those previously formed, with certainty.

Noll (I) proceeded, with this object, with *Caulerpa* and some other marine algæ by producing a precipitate of Berlin blue or Turnbull's blue in the membranes of the plants under investigation without injuring their vitality, and then allowing them to grow more, under favorable conditions. The newly-formed membranes must then, plainly, be colorless ; and those which have grown, perhaps by intussusception, must show a lighter color.

297b. To produce a precipitate of *Berlin blue* in the membranes, Noll (I, 111) placed the algæ first, for one or a few seconds, in a mixture of one part sea-water and two parts fresh water to which was added enough *potassium ferrocyanide* to give the solution the specific gravity of sea-water. Then the algæ were rapidly passed through a vessel of pure sea-water and placed for one half to two seconds in a mixture of two parts sea-water, one part fresh water, and a few drops of *ferric chloride*.* The depth of the coloring is increased markedly by repeating the proceeding several times.

For producing *Turnbull's blue*, which Noll thinks less suitable, he used the corresponding solutions of potassium ferricyanide and ferrous lactate.

297c. But it should be remarked concerning these stainings that they are gradually destroyed, probably by the excretion of alkali. But they can be renewed at any time by placing the algæ in a solution of potassium ferro- (or ferri-) cyanide acidified with pure hydrochloric acid.

Finally it may be observed that, according to Noll's researches, the vitality of the algæ is not destroyed by these manipulations and the precipitate is in this way very uniformly deposited in the ·membranes, provided they present no chemical differences, so that they show the same depth of color in all the layers.

297d. Zacharias (IV, 488) has lately used *Congo red* in the same manner as Berlin blue. ˉ He worked with root-hairs of *Lepidium*, which he placed for 15–30 minutes in a solution of Congo red in water from the public supply and then allowed to grow further in moist air. But, since a decomposition of Congo red takes place in light, the seedlings must be cultivated in the dark.

297e. Congo red was earlier used by Klebs (III, 502) in the investigation of the growth of the membranes of various algæ. This author found that Congo red has the remarkable property of leaving membranes already formed colorless or almost so, while it gives a red color to forming

---

* This solution must be freshly prepared each time it is used, as it decomposes in a short time.

membranes. Klebs used in these studies a .01% solution of Congo red or a suitable culture fluid to which the same proportion of the dye (1 : 10,000) was added. But it should be observed that the Congo red deposited in the membranes strongly hindered their superficial growth in Klebs' experiments, while their growth in thickness was so much the more increased, and the vitality of the cells was in no wise destroyed.

11. The Finer Structure of Cell-walls.

297f. Many cell-membranes, especially those of considerable thickness, are well known to be made up of various lamellæ or layers parallel to their surfaces (*stratification*). In many there occur band-like differentiations within the same layer, which, according to Correns (III, 324), always have a spiral course (*striation*). Finally, one finds not uncommonly radially arranged lamellæ of varying optical properties (*transverse lamellation*).

The observation of these differentiations may in many cases be conducted on the unchanged membranes. But in general they come out much more plainly if the membranes are treated with swelling media; and, besides those mentioned in § 10, chloroïodide of zinc is in many cases very useful.

297g. Three factors may enter into the problem of the cause of the optical appearances described, which have been thoroughly discussed by Correns (III): I. Sculpturing of the wall; II. Differentiation of the wall into strips or layers of unequal water-content with similar chemical constitution; and III. Differentiations of the wall which possess, with similar water-content, unequal refractive power, and therefore depend upon material differences. Besides these, only combinations of these three factors are possible.

297h. *Sculpturing* of the wall may produce especially striation. This then falls into the category of partial thickenings of the membrane, and deserves to be considered here only because, when very delicate, it cannot be distinguished from true differentiations of the membrane without much difficulty, and often accompanies these.

Striation due to sculpturing of the membrane is, plainly, only visible when the membrane and the mounting fluid have different refractive indices. And it becomes the plainer as this difference is the greater, disappearing entirely as the refractive indices become equal. For example, Canada balsam has almost the same optical density as cell-walls; on the other hand, Correns used *methyl-alcohol* with good results, on account of its low refractive index. This is only 1.321 and therefore less than that of water (1.336).

297i. Further, it is evident that it is unimportant in case of differentiations depending wholly on sculpturing of the wall, in opposition to those which are to be referred exclusively to unequal water-content, whether the membranes are placed in a mounting fluid of similar refractive index in a dry or swollen state. But the use of this criterion encounters difficulties, as Correns has shown (III, 260), if rifts or canals in the interior of the membrane are involved, as in the bast-cells of *Nerium*, where the different layers have different systems of striation. In this case it does not seem practicable to fill these capillary spaces with ethereal oils or with balsam without the removal of imbibed water. But even in this case, the behavior of the dried membranes on being imbedded in Canada balsam or the like may permit positive conclusions as to the nature of the differentiations in question, since only a slow expulsion of the enclosed air from capillary spaces in the interiors of membranes can take place.

For distinguishing water-bearing clefts from substances rich in water, chloroïodide of zinc and various dyes may be used. Clearly, the capillary spaces must always stand out as colorless streaks on suitable sections, while the parts richer in water may show a more or less deep stain.

297k. Differentiations due to *unequal water-content* must, plainly, disappear on drying, as a rule. The presence of such differentiations may therefore be recognized by examining the objects in the same anhydrous mounting fluid (such as Canada balsam), a part dry and a part moist. But it should be observed that the complete removal of water can only be accomplished by drying at a temperature of 50° to 100° C., according to the nature of the object. The dehydrating media used by various authors for the same purpose, especially absolute alcohol, do not give demonstrative results (cf. Zimmermann I, 87).

297l. It is also to be noticed that, where the water-content is unequal, changes in form must occcur on drying, and therefore, as Correns (III, 262) has specially observed, a certain distinction between differentiations due to sculpturing of the wall and those due to unequal water-content cannot always be drawn from the comparison of dried and moist membranes. Concerning the possibilities in this respect, Correns' work (III) may be consulted.

297m. The presence of differences in water-content was demonstrated by Correns (III, 294) by impregnating the membranes with a salt-solution (NaCl), which is not recognizably accumulated. The conclusion is then justified that where the salt occurs in greater quantity this is in consequence of greater water-content. Correns used for this purpose potassium ferrocyanide and silver nitrate, and made the salt contained in the walls visible by conversion into a colored precipitate.

Concerning the method of using the *potassium ferrocyanide*, it may be observed that Correns placed the membranes, previously washed in water and then dried by warming to 50° to 100° C., for a few minutes in a 10% solution of the substance and then placed them, after superficial drying on filter-paper, but without washing, in a dilute solution of ferric chloride, in which the formation of Berlin blue at once takes place. In the bast-cells of *Nerium* or *Vinca*, which are especially suited to these experiments, the striation becomes clearly visible with this treatment, even after drying and mounting in cedar-oil and Canada balsam.

Correns (III, 295) was able to show that the potassium ferrocyanide is merely absorbed, but not accumulated, by placing dry starch-grains in a solution of this salt, and testing its concentration before and after. It was thus shown that no marked changes in concentration were produced by the absorption of the starch. It had previously been shown by Sachs that potassium ferrocyanide rises with the same rapidity as the water in a strip of filter-paper.

297n. For "*silvering*," Correns used essentially the methods which have been employed for a long time in animal anatomy. He placed the previously well-dried objects first in a 2–5% solution of *silver nitrate*, and then, after superficial drying, but without washing them, in a .75% solution of salt (NaCl). The silver chloride thus precipitated in the membranes is best reduced by light, for which a few hours in direct sunlight are sufficient. The objects are then dried and mounted in Canada balsam. There may then be observed in the parts rich in water a strong blackening due to the reduced silver, which also forms, in part, small opaque granules. Here again the bast-cells of *Nerium* and *Vinca* are to be recommended, but they should first be washed some time with water for the removal of silver-reducing substances.

According to Correns (III, 296), silver nitrate is accumulated to a slight extent. A slight change in the concentration of the solution is produced by starch, and the silver salt remains somewhat (about $\frac{3}{20}$) behind the water in filter-paper.

297o. Differentiations of the membrane due to *chemical differences* are especially recognizable by the fact that they are to be seen in the membranes whether dry or full of water. Suitable objects for the study of this group of differentiations are furnished by the large pith-cells of *Podocarpus elongatus* and other species (cf. Zimmermann I, 149).

The so-called transverse lamellation of the bast-cells is due partly to chemical differences and partly to unequal water-content, according to Correns (II, 298). The chemical difference is shown by the fact that the radial, more strongly refractive lamellæ remain unstained in a pretty concentrated solution of *methylene blue*, while the rest of the substance of the wall is deeply stained by it. The stronger refractive power is partly destroyed by Schulze's macerating mixture, evidently by the solution of the substance which causes the stronger refraction; and correspondingly, the membranes so treated are evenly stained throughout by methylene blue.

297p. Finally, the *carbonization* or *pulverization* methods introduced by Wiesner (V) into botanical microscopy may be here described. By these the vegetable cell-wall is broken up into filamentous and then into spherical bodies, which are called by this author *dermatosomes*, but whose significance cannot be discussed here.

Linen fibres may serve as suitable objects for the trial of pulverization methods. These are laid, according to Wiesner (III, 14), for 24 hours in a 1% solution of hydrochloric acid, then freed from adhering fluid and warmed to 50° or 60° C., until the substance is completely dry, which can be accomplished in 30 to 50 minutes if small quantities of fibre are used. The fibre may then be broken into an extremely fine powder by gentle pressure.

With other objects a longer stay in hydrochloric acid, or the use of higher temperatures for drying, is necessary. Wiesner accomplished the pulverization of endosperm-cells of *Phytelephas* only after the action of hydrochloric acid for months.

Pfeffer (IX) has lately shown that the carbonization methods lead to the same results with artificially prepared collodion membranes.

B. The Protoplasm and Cell-sap.

298. Conclusions concerning the morphological characters of the protoplasm have been sought for not only by direct observation of living material, but also by means of microchemical reactions and staining methods. Although it would seem probable *a priori* that microchemistry would prove an aid of the first importance, it has not yet justified these expectations; which is due largely to the fact that macrochemical studies of the structures in question have been carried through with any exactness in very few cases, since

they have to overcome great difficulties in the small size
and ready decomposition of the bodies concerned. Therefore it is not yet certain whether the various organs of the
plasma-mass, like the nucleoli or the leucoplasts, consist
always of the same or only of related chemical compounds;
although their behavior with certain staining agents makes
the former supposition seem probable for many cases. For
we possess certain staining methods which act with such
precision as to render them worthy of places with the best
microchemical reactions, and to assure the first place in
the investigation of protoplasmic structures to staining
methods. But it cannot be doubted that in the immediate
future microchemical reactions may be of the greatest
importance in the study of the plasma-body.

In the following pages we will first discuss the methods of
recognizing the various inclusions and differentiations of
the protoplasm, and then describe some methods which
have been used in the study of various general characteristics of the plasma-mass and the cell-sap, and of certain
physiological processes.

1. The Nucleus and its Constituents.

299. The advances which our knowledge of the morphological characters of the nucleus have made in recent decades are almost exclusively due to staining methods, which
makes explicable the fact that the most various natural and
artificial dye-stuffs have been tested as to their applicability
in this respect, and that innumerable staining methods have
been most warmly recommended by their discoverers.

It cannot be necessary for me to enumerate all these
methods here, but rather to limit myself to the best of them,
which are capable of general application and do not fail in
difficult cases. I begin with the enumeration of the most
important fixing and staining methods, and add some general remarks as to the staining of the nucleus in various
cells and on the recognition of its various constituents.

I. The various Methods, in general.

a. Fixing Methods.

α. Alcohol, C_2H_5OH.

300. Exposure for 24 hours is usually sufficient for fixing nuclei, but a longer action does no harm.

If one desires to prepare sections free-hand from alcoholic material, it is often useful to place the material for 24 hours previously in a mixture of equal volumes of alcohol and glycerine, or of alcohol, glycerine, and water, which makes them much better adapted to cutting.

Concerning the addition of sulphurous acid to such preparations as blacken in pure alcohol, see § 34.

β. Iodine.

301. Iodine has been chiefly used for fixing in an aqueous solution of iodine and potassium iodide. Berthold (II, 704, note) recommends for marine algæ a concentrated solution of iodine in sea-water, which may be prepared by the addition of a few drops of alcoholic iodine solution to pure sea-water. According to Berthold, it is sufficient to move the algæ about in this fluid from half a minute to a minute. They are then transferred directly to 50% alcohol and, if this fluid be changed a few times, can be placed in the staining fluid in a few minutes.

Overton (I, 530) has used the vapor of iodine, which may easily be obtained by warming iodine-crystals in a narrow test-tube, for fixing, with the advantage that they may be entirely expelled by gentle warming (to 30° or 40° C.) and require no washing out of the fixing medium. Their use is especially to be recommended for small objects (cf. § 40).

γ. Bromine and Chlorine.

302. The vapor of bromine is recommended by Strasburger (I, 399) for fixing *Fucus*. [Zimmermann (VIII) recommends chlorine gas for fixing such algæ as *Cladophora* and *Zygnema* without contraction of the protoplasm.]

δ. Picric Acid, $C_6H_2.(NO_2)_3.OH$.

303. Picric acid is used mostly in a concentrated aqueous or alcoholic solution. Its action for 24 hours is generally sufficient. Before staining it must be carefully washed out, for which purpose running water is especially useful (cf. § 35). But in many cases it is better to wash with alcohol, in which picric acid is more readily soluble than in water.

ε. Picro-sulphuric Acid.

304. Picro-sulphuric acid may be prepared, according to the recipe proposed by Mayer, by mixing 100 volumes of water and two of concentrated sulphuric acid, then shaking up with it as much picric acid as will dissolve, and finally diluting the whole with three times its volume of water. Picro-sulphuric acid has the advantage over pure picric acid that it is more easily washed out. It has been much used with the lower organisms.

ζ. Chromic Acid, H_2CrO_4.

305. Chromic acid has been used with the best results for fixing algæ, especially in a 1% aqueous solution. Its action for a few hours is always sufficient for these plants; but with larger tissues it is better to allow the medium to act for 24 hours. Before staining, the chromic acid must always be well washed out, for which running water is best (§ 35). Overton recommends (I, 10) for this purpose a weak aqueous solution of *sulphurous acid*. This makes objects fixed in chromic acid fit for staining with hæmatoxylin and carmine in a few minutes.

Chromic acid has the disadvantage of often causing, especially in tissues rich in tannin, the formation of precipitates which hinder observation.

306. Finally, it may be observed that, according to Virchow (I), objects fixed with chromic acid should be brought in contact with alcohol only in the dark, before the complete removal of the acid, since their power of staining is lessened by the formation of a precipitate in the light. In the alcohol used for washing, a precipitate is also formed in the light; but when this is removed by filtering the fluid can again be used for washing.

η. Chrom-formic Acid.

307. Rabl recommends (I, 215) a mixture of 200 grams of $\frac{1}{2}\%$ chromic acid with four or five drops of concentrated formic acid for fixing nuclear figures. It must be freshly prepared before use, each time, and should act for 12–24 hours. It must be well washed out with water before staining.

θ. Osmic Acid, OsO_4.

308. To fix sections or small objects, the vapor of osmic acid may best be used by placing the objects in a drop of water on a cover-glass or slide and bringing the drop over the mouth of a bottle containing a 1% or 2% solution of osmic acid. Killing and fixing take place almost instantly. For fixing larger pieces of tissue 1% osmic acid may be used and may act for several hours without harm.

Osmic acid, which is indisputably one of our best fixing media, has the disadvantage of producing brown or black precipitates with very various substances. But in most cases these precipitates may be removed subsequently without injury to the protoplasmic structure, and for this use *hydrogen peroxide* is best adapted. Overton (I, 11) recommends for this purpose a mixture of one part of commercial peroxide with 10–25 parts of 70–80% alcohol. I have also observed that, even after the use of the concentrated commercial solution of peroxide, which decolorizes at once, especially on gentle warming, the karyokinetic figures remain quite sharp and unchanged in microtome sections.

ι. Chrom-osmic-acetic Acid.

309. Mixtures of chromic acid, osmic acid, and acetic acid have been used with the best results in the study of the karyokinetic figures, especially by Flemming. This author has used solutions of very different strengths; but the following, which have both proved very good with plant-cells, may be mentioned here. The first, *dilute mixture* contains .25% of chromic acid, .1% of osmic acid, and .1% of acetic acid. The more *concentrated mixture* is prepared from 15 volumes of 1% chromic acid, 4 volumes of 2% osmic acid, and one volume (or less) of glacial acetic acid.

Both mixtures must be carefully washed out with water. Blackening due to the osmic acid may be removed with hydrogen peroxide (cf. § 308).

For preserving objects fixed with the above mixture, Flemming (III, 687, note) recommends a mixture of water, alcohol, and glycerine, in about equal parts. This affects their staining capacity less than pure alcohol.

κ. Corrosive Sublimate, $HgCl_2$.

310. Corrosive sublimate is usually best used in concentrated alcoholic solution, though the concentrated aqueous solution often does well. Action for a few hours is always sufficient for complete fixing, but it may be left on the objects without harm for 24 hours. Alchohol to which enough *iodine* has been added to give a dark brown solution may be used for washing out the sublimate. If it is not wholly washed out, needle-shaped or sphærite-like crystals of sublimate may be seen in the preparation and may easily deceive beginners. But where the sublimate has not been wholly removed before imbedding in paraffine, it may subsequently be washed out, even from microtome sections, with the iodine-alcohol.

If it is desirable to avoid alcohol in washing out sublimate, the mixture, proposed by Haug (I, 13), of two parts tincture of iodine, one part potassium iodide, 50 parts glycerine, and 50 parts water may be used. It should be renewed until no further decolorization occurs.

I will remark that objects fixed with sublimate must not be touched with iron forceps or the like before it is wholly washed out, since globules of mercury are thus easily produced within the objects. In this case forceps with platinum or horn points may be used.

λ. Platinum chloride, $PtCl_4$.

311. A ⅓% aqueous solution of platinic chloride has been recommended by Rabl (I, 216) for fixing nuclear figures. Its use brings out especially the longitudinal splitting of the segments of the nuclear thread and the chromatin spheres. It should act, in general, for 24 hours.

μ. Chromic-acid-Platinum-chloride.

312. Merkel used the following combination: 100 volumes of 1% chromic acid, 100 volumes of 1% platinic chloride, and 600 volumes of water. This medium renders good service also with vegetable objects. It should act for 24 hours.

ν. Platinum-chloride-Osmic-acetic acid.

313. F. Hermann (I, 59) has recommended the following mixture: 15 volumes of a 1% platinic chloride solution, one volume of glacial acetic acid, and two or four volumes of 2% osmic acid. Hermann allows this mixture to act from one to two days. To show the achromatic nuclear figure, he washes the fluid out in running water, hardens in alcohols of increasing strength, and then lays the objects in crude pyroligneous acid for 12 to 18 hours (cf. Hermann, II, 571).

b. Staining Methods.

α. Hæmatoxylin.

314. Besides carmine, hæmatoxylin has been most used for staining nuclei, and we have a great number of recipes for the preparation of specially active hæmatoxylin solutions. Among these the so-called *Delafield's hæmatoxylin* solution, also often erroneously called *Grenacher's* hæmatoxylin, seems to deserve preference in most cases. It is prepared as follows: 4 grams of hæmatoxylin are dissolved in 25 ccm. of alcohol, then 400 ccm. of a concentrated aqueous solution of ammonia alum are added, and the mixture is allowed to stand in the light for 3 or 4 days and is filtered; 100 ccm. of glycerine and 100 ccm. of methyl alcohol are added, the whole is allowed to stand again for a

few days and is filtered again. Since the method of preparing it is somewhat complicated, many will prefer to obtain it ready prepared from a chemist (e.g., from Dr. G. Grübler, Leipzig).

315. The so-called *Böhmer's hæmatoxylin* is also very useful. It is prepared from a concentrated alcoholic solution of hæmatoxylin which contains .35 gram of hæmatoxylin to 10 grams of alcohol and will keep indefinitely. A few drops of this are added to a solution of .10 gram of alum in 30 ccm. of water. This mixture is allowed to stand for a few days and is filtered before use.

P. Mayer (III) obtained an hæmatoxylin solution that may be used at once by dissolving 1 gram of *hæmateïn* or *hæmateïn-ammonia* in 50 ccm. of 90% alcohol by warming, and then adding the whole to a solution of 50 grams of alum in a litre of water. This solution may be diluted with distilled water for staining, as desired.

Concerning the other solutions of hæmatoxylin which may be valuable in special cases, and may in part be obtained ready for use from various chemists, reference may be had to the compilation of Gierke (I, 32-35).

316. If it is desired to stain sections with hæmatoxylin, they are best placed in a very dilute solution and left in it for a considerable time (1 to 24 hours). With alcoholic material it is advisable to place it in water for a short time before staining it, as otherwise precipitates are readily formed.

Beautifully differentiated stainings may usually be obtained by staining sections too deeply ("overstaining") and then washing them out with a suitable fluid. With an hæmatoxylin stain, a solution of *alum* (about 2%) is commonly best; but it must be thoroughly washed out before the transfer to alcohol or to Canada balsam, as otherwise alum crystals will be formed in the preparation.

Acid alcohol has also been recommended for washing out hæmatoxylin; but the acid must be completely removed with pure alcohol before the final mounting.

Very good nuclear staining may often be obtained by

placing objects stained with hæmatoxylin for a short time in a 1% solution of *potassium bichromate* or a concentrated aqueous solution of *picric acid*. Both fluids must, of course, be carefully washed out before the transfer to Canada balsam or glycerine-gelatine.

317. In dealing with objects to be sectioned with the microtome, very pure nuclear stains may usually be obtained by staining the objects *in toto* ("staining in mass") before imbedding in paraffine. Since hæmatoxylin cannot penetrate the cuticle and therefore only penetrates from cut surfaces, very different depths of staining are obtained, and, at some distance from the original surface, only a nuclear stain. Large objects must sometimes be left in the staining fluid, which in this case should be used pretty dilute, for some time (often several days), in order to be sufficiently stained. Very good mass-staining may be obtained by first staining large pieces of tissue deeply with hæmatoxylin and then placing them for a considerable time in a 1% solution of potassium bichromate.

β. Carmine.

318. Only a few of the numberless different carmine solutions can be described in detail here. These, as well as various other staining media containing carmine, can be obtained ready for use from chemists.

1. *Grenacher's borax-carmine* can be prepared by dissolving 4 grams of borax and 2 to 3 grams of carmine in 93 ccm. of water, then adding 100 ccm. of 70% alcohol, shaking and filtering. This solution is used for staining in mass as well as for sections. For washing, acid alcohol and a solution of borax or oxalic acid in spirit are recommended.

2. *Beale's Carmine.*—.6 gram of carmine is shaken up with 3.75 grams of *liquor ammonii caust.* [aqua ammoniæ, U. S. P.], then boiled for a few minutes; after an hour, 60 grams of glycerine, 60 grams of water, and 15 grams of alcohol are added, and the whole is finally filtered.

3. *Ammonium carminate* is best prepared by dissolving in water to which a little (about .5%) ammonium carbonate has

been added, the commercial dry ammonium carminate (the so-called Hoyer's ammonium carminate). Alcohol or acid alcohol is best used for washing.

4. *Sodium carminate* can be obtained in solid form, and may be dissolved in an aqueous .5% solution of ammonium carbonate.

5. *P. Mayer's carmine solution* is prepared by rubbing up 4 grams of carmine in 15 ccm. of water, then adding 30 drops of hydrochloric acid while warming, and finally adding 95 ccm. of 85% alcohol, boiling, neutralizing with ammonia, and filtering when cold. Is used both for staining sections and for staining in mass.

6. *Carminic Acetate.*—Ammonium carminate is decomposed with acetic acid added drop by drop in the least possible excess until the cherry-red fluid has become brick-red, when it is filtered. For washing, a mixture of one part hydrochloric acid in 200 parts glycerine or one of one part formic acid in 100 parts glycerine is recommended.

7. *Picrocarmine* is the term applied to variously prepared mixtures of picric acid and carmine. Only the simplest recipe (Hoyer's) need be given here. According to this, pulverized carmine is dissolved in a concentrated solution of neutral ammonium picrate.

P. Mayer (III) now uses pure *carminic acid* for the preparation of carmine solutions, of which he especially recommends the following:

8. *Carmalum* is prepared by dissolving 1 gram of carminic acid and 10 grams of alum in 200 ccm. of distilled water, with heat. The solution may be decanted off or filtered. To protect it against decomposition, this author finally adds a few crystals of thymol, or .1% of salicylic acid or .5% of sodium salicylate. On washing with water the protoplasm remains somewhat colored. To obtain a purely nuclear stain, the washing must be carefully done with a solution of alum or a weak acid.

9. *Paracarmine* is prepared according to the following recipe: 1 gram of carminic acid, ½ gram of aluminium chloride, and 4 grams of calcium chloride are dissolved in 100

ccm. of 70% alcohol with or without heat, the whole is allowed to stand and then filtered. Washing with acid alcohol is usually unnecessary, but a weak solution of aluminium chloride in alcohol, or alcohol containing 2½% of glacial acetic acid, is sufficient for all cases.

According to P. Mayer's statements, only Grenacher's borax-carmine, of the numberless solutions heretofore recommended, presents any advantages over carmalum and paracarmine.

A. Meyer (V) recommends especially for staining the nuclei of pollen-grains:

10. *Chloral Carmine.*—This is prepared by heating for 30 minutes on the water-bath .5 gram of carmine, 20 ccm. of alcohol, and 30 drops of officinal hydrochloric acid,* and then adding 25 grams of chloral hydrate. After cooling, the solution is filtered. It stains the nuclei of pollen-grains deep red in ten minutes. In many cases I have obtained good nuclear stains with this medium by washing the sections in boiling water on removing them from it. Such preparations may be mounted in glycerine or may be transferred in the usual way to Canada balsam.

[11. *Czokor's Alum cochineal* has proved so good as a nuclear stain for many vegetable tissues that it should have a place here. It is prepared as follows: 7 grams of crude pulverized cochineal (the dried insects) is boiled in a solution of 7 grams of burnt alum in 700 ccm. of water until the whole is reduced to 400 ccm. The solution is then carefully filtered and a few crystals of carbolic acid (phenol) are added as a preservative.]

319. All these solutions of carmine are adapted as well for staining in mass as for staining sections. But they generally penetrate pretty slowly and require to act for a rather long time. They are inferior in the sharpness of their staining to many other dyes, in most cases; but usually give a very purely nuclear stain. The cell-wall is especially seldom stained by carmine.

[* This has a specific gravity of 1.13 or 17° B.]

γ. Safranin.

320. The aniline-water-safranin recommended by Zwaardemaker (I) is best used for staining with this dye. It may be prepared by mixing equal volumes of a concentrated alcoholic solution of safranin and aniline-water*; and should be allowed to act for an hour or longer. Preparations may be washed with alcohol or acid alcohol, i.e., alcohol containing about .5% of hydrochloric acid.

δ. Gentian Violet (Gram's Method).

321. The method originally proposed by Gram for staining isolated bacteria (cf. § 470) is in many cases well adapted for staining nuclei, especially in material which has been fixed with one of the acid mixtures recommended in §§ 309, 312, and 313. It is well to use here as a staining fluid, a mixture of 3 grams of aniline, 1 gram of gentian violet, 15 grams of alcohol, and 100 grams of water. The sections remain in this one or a few minutes, the stain is washed off with alcohol, and a solution of 1 part of *iodine* and 2 parts of potassium iodide in 300 parts of water is at once added. This gives the sections a dark color and is then washed off with alcohol; clove-oil is added, which extracts more coloring matter from the sections, and usually first brings out the characteristic differential staining; finally the sections are mounted in Canada balsam.

It may be especially remarked here that in this process the action of the clove-oil is in no way merely a clearing one, as has often been said. If it be replaced by xylol, even after thorough washing with alcohol, not nearly so good nuclear stains will be obtained, in many cases, as where clove-oil is used.

Finally, a very good double stain may be obtained by dissolving *eosin* in the clove-oil used for washing. The walls and the achromatic nuclear figure appear pure red, and the chromatic figure violet.

---

* This is prepared by shaking water with an excess of aniline, and contains about 3.5% of aniline.

ϵ. Safranin and Gentian Violet.

322. According to the methods recommended by Hermann (I, 60), the sections are first placed for 24-48 hours in a solution of safranin which contains 1 gram of the dye to 10 ccm. of alcohol and 90 ccm. of aniline-water. They are then treated successively with water, acid alcohol, and alcohol, but so that they remain still too deeply stained for direct observation. From the alcohol, the sections are then transferred for 3 to 5 minutes to a solution of gentian violet containing 1 gram of that dye to 10 ccm. of alcohol and 90 ccm. of aniline-water. The sections are then quickly rinsed off in alcohol and placed in a solution of iodine and potassium iodide prepared as for Gram's method (§ 321), in which they remain for one to three hours until they are quite black. Then they are differentiated in alcohol, cleared in xylol, and mounted in Canada balsam.

In successful preparations the nucleoli of resting nuclei are vivid red, the nuclear framework blue-violet. Of the karyokinetic figures the spirem and dispirem are blue, while the intermediate stages are red. The achromatic figure is lightly stained yellow-brown by the iodine.

ζ. Safranin-Gentian-violet-Orange.

323. According to Flemming (III, 685, note), objects fixed with chrom-osmic-acetic acid (cf. § 309) or with Hermann's platinum-chloride-osmic-acetic acid (cf. § 313) may best be placed for 2 or 3 days in a concentrated alcoholic solution of safranin which is diluted with about an equal volume of water and a little aniline-water. They are then washed in water and then extracted with alcohol containing at most .1% of hydrochloric acid, or with pure alcohol. After a brief washing in water, the objects are placed for 1 to 3 hours in a concentrated aqueous solution of gentian violet, and then, after another short washing in water, in a concentrated aqueous solution of orange,* in which its dark color is gradually dissolved out. After a few minutes, while blue clouds of

---

* This may be obtained from Dr. G. Grübler under the name "Orange G."

color are still rising, the objects are placed in neutral absolute alcohol, and, after the repeated renewal of this, in clove-oil or bergamot-oil, which still extracts light clouds of color, and finally mounted in balsam. The difficulty of this method consists in determining the right minute when the objects in alcohol and in oil are neither too much nor too little washed out.

In successful preparations the chromatin will be purple-red, the threads of the achromatic spindle gray-brown, gray, or violet, the centrosomes (cf. § 348a) the same or light reddish.

324. I have tested the last two methods upon the root-tips of *Vicia Faba* and on various other vegetable objects and can most heartily recommend them for staining vegetable nuclei. I obtained the most instructive preparations by leaving the sections (I worked with microtome sections) from $\frac{1}{2}$ to 2 hours in Hermann's aniline-water-safranin. I then washed them successively in acid alcohol and in alcohol, added aniline-water-gentian-violet, which I left on the sections for 2 to 4 minutes at most, washed off the dye with water, immersed them for 5 minutes or longer in Gram's iodine solution, and then washed either with alcohol or successively with alcohol and clove-oil, and mounted in Canada balsam ; or I also subsequently stained with orange. In the latter case I washed the sections, after they came from gentian violet, but a very short time in alcohol and then placed them in a concentrated aqueous solution of orange, which was allowed to act a few minutes, then washed them in alcohol and transferred, in the usual way, to Canada balsam.

In good preparations where orange was used, the nucleoli and the karyokinetic figures from aster to dyaster were deep red, the nuclear framework of resting nuclei and the spirem and dispirem were violet or blue, the achromatic nuclear figure and the cytoplasm were yellow-brown.

[Gjurasin has used these methods with much success, after many others had failed, for bringing out the karyokinetic figures in the asci of *Peziza*, after the material had been fixed for two days in Flemming's chrom-osmic-acetic acid.]

η. Fuchsin.

325. The sections are first placed for 15 minutes or longer in a concentrated aqueous solution of fuchsin, then covered with a concentrated solution of *picric acid* in 2 parts water and 1 part alcohol, in which the stain becomes dark violet, then washed with 90% alcohol as long as any color is given off, then quickly rinsed in absolute alcohol, transferred to xylol and finally to balsam. A very deep nuclear stain is thus obtained.

This stain may be combined with *methyl blue* by placing the sections for about a quarter of an hour in an aqueous solution of methyl blue, after washing out the picric acid with alcohol. The methyl blue is then washed off and the sections are mounted, as usual, in balsam.

ϑ. Fuchsin-Methyl-green.

326. Guignard (I, 19) recommends for nuclear staining an aqueous solution of fuchsin and methyl green which contains enough of the two dyes to make it appear deep violet. The solution may be very feebly acidified with acetic acid. It very quickly stains the nucleus blue-green and the protoplasm bright red. In case of overstaining, it may be washed out with water.

ι. Fuchsin-Iodine-green.

327. According to Strasburger (I, 575), a mixture of fuchsin and iodine green, proposed by Babes, is very useful for vegetable objects. It is prepared by pouring a solution of iodine green in 50% alcohol into an open dish and adding fuchsin, also dissolved in 50% alcohol, until the fluid takes a markedly violet color. The sections to be stained remain about a minute in this solution and are then transferred to glycerine.

c. Simultaneous Fixing and Staining.

α. Methyl-green-Acetic-acid.

328. To obtain a rapid staining of the nuclei with living objects, the solution of methyl green in 1% acetic acid, recommended especially by Strasburger, may often be used with good results. But very weak solutions of the dye must generally be used. Examination may be made in the stain itself or in glycerine. The preservation of preparations stained with methyl green is not possible.

β. Picro-nigrosin.

329. The solution of nigrosin in a concentrated aqueous solution of picric acid, recommended by Pfitzer (I), may be used with good results for simultaneous fixing and staining, especially with algæ. In order to remove the chlorophyll at the same time, a solution of nigrosin and picric acid in 96% alcohol may be used. At least 24 hours' time is usually necessary for good staining. The preparations are then washed in glycerin and may be preserved in glycerine-gelatine. But the stain comes out much more finely in balsam or the like. It is preserved well in either medium.

d. Staining intra vitam.

330. The staining of living nuclei was first accomplished in the higher plants by Campbell (I) by means of *dahlia*, *methyl violet*, and *mauveïn*. The author mentioned placed pieces of the objects to be studied in a .001% to .002% solution of one of these dyes, and usually left them there for several hours. The stamen-hairs of *Tradescantia virginica* are recommended as especially adapted for these studies, and Campbell succeeded in staining their nuclei while in process of division. But these stainings are usually pretty faint and always much less differentiated than good stainings of fixed nuclei. This staining *intra vitam* has not led to any important results concerning the morphology of the nucleus.

II. The Resting Nucleus and its Constituents.

a. Recognition of the Resting Nucleus.

331. If one is concerned simply with the recognition of resting nuclei in an organ or a tissue system, this should no longer present serious difficulties in the Phanerogams, the Pteridophytes, or the Mosses. Especially by fixing with chrom-osmic-acetic acid (§ 309) or with chromic-acid-platinum-chloride (§ 312), and staining according to one of the methods described in §§ 320 to 325, this object may always be realized certainly and without great trouble. By staining in mass with hæmatoxylin (§ 314) I have also always obtained very sharp nuclear staining at a little distance from the surface. Very instructive preparations may also be usually obtained from algæ and fungi by the use of the same methods. For most algæ 1% chromic acid (§ 305) is an excellent fixing medium, and good stainings may also be obtained by the picro-nigrosin method (§ 329).

332. To stain the nucleus in the starch-filled oöspheres of *Nitella*, Overton (II, 35) used a mixture of *potassium ferrocyanide* and hydrochloric acid, diluted with 8–10 times its bulk of water. The starch was then converted into sugar by the hydrochloric acid and, at the same time, the Berlin blue produced by the decomposition of the ferrocyanide stained the nucleus. This author recommends chloral hydrate for clearing.

333. Special difficulties are sometimes due to the cuticularized or slightly pervious membranes which often invest reproductive organs. In such cases the subsequent staining of microtome sections will give the best results.

334. If one wishes, in especially difficult cases, as in following the development of spermatozoids, to obtain a very sharp differentiation of nucleus and cytoplasm, double staining may be used with good results. Guignard (I) used for this purpose, with material fixed with osmic acid or with alcohol, the staining method with fuchsin and methyl green described in § 326.

b. The Constituents of the Resting Nucleus.

335. If we examine more closely the various constituents of the nucleus, the *nucleolus* presents the part which stains most deeply with most staining methods. In many objects, after the use of the above-described methods and very thorough washing, only the nucleoli appear very deeply stained in the resting nuclei. On the other hand, the so-called *chromatin spheres* of the nuclear framework are distinguished by marked staining power, and great differences in this respect occur in different plants and tissues, as well as in similar cells of different ages; so that it often happens that only the nucleoli are stained in the younger parts, and only the nuclear framework in the older ones, by the same method. The cause of these differences cannot yet be stated; but it is not improbable that bodies of various sorts are contained in the so-called nuclear framework.

336. The most certain distinction between chromatin-globules and nucleoli may be reached by means of the Hermann-Flemming safranin-gentian-violet methods (cf. §§ 322–324), which give a beautiful red color to the nucleoli and a violet blue to the chromatin-globules, especially with material fixed with chrom-osmic-acetic acid (§ 309) or with platinum-chloride-osmic-acetic acid (§ 313). Whether these methods give as clear results in all cases must be determined by further studies.

336a. The general distribution of *erythrophilous* and *cyanophilous* constituents of the nucleus was first recognized by Auerbach in animal cells. These studies have lately been extended to plants by Rosen (I), who recommends the following methods:

1. *Acid Fuchsin-Methylene Blue.*—The sections are first stained with Altmann's acid fuchsin (cf. § 345), then washed successively with picric-acid-alcohol and with water, then stained with methylene blue, soaked in alcohol, and mounted in balsam.

2. *Fuchsin and Methylene Blue.*—First stain with an aque-

ous .1% solution of fuchsin, then wash with water, stain again with .2% aqueous solution of methylene blue, and finally wash with alcohol or with a mixture of three parts xylol and one part alcohol.

3. *Acid Fuchsin and Methylene Blue.*—The microtome sections fastened to the slide are stained for half an hour in a .1% aqueous solution of acid fuchsin, then quickly rinsed with water and treated for $\frac{1}{2}$–1 minute with a .2% aqueous solution of methylene blue. The superfluous stain is then removed with alcohol and the preparation, as soon as it is air-dry, is extracted with clove-oil, in which it may be left from 6 to 24 hours. The clove-oil is then washed out with alcohol or xylol-alcohol, and the preparation finally mounted in balsam. This method has also been used by Schottländer (I). He states, however, that the time of exposure to the methylene blue must be much varied (between a few seconds and two minutes).

337. In many cases *digestive fluids* may be used for the recognition of the chromatin-globules. These are not attacked by pepsin, but are quickly dissolved by *trypsin* (cf. §§ 232–4).

338. I will remark here that Altmann (III) has observed a granula-structure in the resting nucleus, which cannot be further discussed here, since he has not published the methods used by him.

339. The *nuclear membrane* is also slightly capable of staining. According to Fr. Schwarz (I, 123), it is best made visible by a 20% solution of common salt or of mono-potassium sulphate, or by a concentrated solution of potassium bichromate or the mixture of potassium ferrocyanide and acetic acid mentioned in § 238.

III. *The Karyokinetic Figures.*

340. In nuclei in process of indirect or karyokinetic division there may be distinguished a *chromatic* and an *achromatic* figure. The former generally consists of a nuclear thread composed of globules or disks arranged in series (chromatin-globules), which first falls into segments, each of which is then split lengthwise. The half-segments thus produced then separate so that one of each pair goes to each daughter-nucleus. The achromatic figure consists, on the other hand, of a number of very delicate threads which are usually arranged in the form of a spindle, so that they are called the nuclear spindle (cf. Fig. 36).

341. The *chromatic figure* has, as the name indicates, a high staining capacity, and is also usually much more deeply stained than the nuclear framework of the resting nucleus. It agrees more with the nucleoli in its behavior with staining media, although it does not in any case originate from them.

In most cases, a very deep staining of the chromatic nuclear figure is obtained by fixing with the concentrated Flemming's chrom-osmic-acetic acid mixture (cf. § 309) and staining with aniline-water-safranin (§ 320). In consequence of their deeper staining, the dividing nuclei are easily distinguished from the resting nuclei, even with pretty low powers. Besides the above, chromic-acid-platinum-chloride (§ 312) and chrom-formic acid (§ 307) are especially useful for fixing, and Gram's method (§ 321) and the fuchsin-picric-acid method (§ 325), for staining the chromatic figure. In most cases very good preparations of the karyokinetic figures are obtained by fixing with alcohol (§ 300) or alcoholic corrosive sublimate (§ 310) or picric acid (§ 303) and staining according to one of the methods mentioned, or with one of the numerous hæmatoxylin (§§ 314–317) or carmine solutions (§§ 318, 319).

I will remark here that often the karyokinetic figures may be made clearly visible without previous fixing by placing sections of living tissues directly in a concentrated aqueous solution of *chloral hydrate*, in which usually only the nucleus is left, of all the cell-contents, and the chromatic figures of dividing nuclei come out with especial sharpness. In this way I have obtained very instructive preparations from young fern-fronds.

342. On the other hand, it is in most cases difficult to make the so-called *achromatic figure* clearly visible. It is, to be sure, capable of staining to a certain degree, especially when hæmatoxylin is used, or on double staining with fuchsin and methyl blue (§ 325), when the chromatic figure is colored red, the achromatic, blue. But, even in these preparations, it is difficult to clearly distinguish the separate threads of which the achromatic spindle is made up; and in difficult cases, as in many fungi, it is usually impossible to bring out this figure in this way. A preliminary treatment with reagents will better produce this result.

[Strasburger recommends fuming hydrochloric acid for bringing out the structure of the "kinoplasm," which includes the achromatic figure.]

How far the methods lately recommended by Flemming and Hermann for animal objects (cf. §§ 322–324) will prove of value in these cases must be determined by further researches (compare also § 348 a–e).

IV. *The Inclusions of the Nucleus (Protein Crystalloids).*

343. As recent investigations have shown, protein crystalloids are pretty widely distributed in the nuclei of the Pteridophytes and Angiosperms (cf. Zimmermann II, 54, and III, 112); but they do not belong to their constant constituents, and it therefore seems to me better to treat them as inclusions of the nucleus, like the starch-grains and crystalloids contained in the chromatophores. No other heterogeneous inclusions have yet been recognized in the nucleus.

344. In the recognition of protein crystalloids their regular crystalline form is in many cases an aid. Thus the crystalline structure of the crystalloids from the leaves of *Melampyrum arvense* and *Candollea adnata*, shown in Fig. 34, 2 and 4, can hardly be questioned. Besides, one very often finds also needle-like or rod-like crystalloids, as in the ovules of

Mimulus Tillingi (Fig. 34, 6), or variously curved forms, as in the wall of the ovary of *Campanula trachelium* (Fig. 34, 1). Finally, bodies of the same chemical relations are widely distributed, which vary little or none from the globular form, as for example in the leaf of *Lophospermum scandens* (Fig. 34, 3). In such cases the certain proof of their crystalloid nature, and especially the distinction between them and the nucleoli, is not possible by the simple examination of living material. But this distinction may be easily and certainly made by the aid of suitable staining methods.

FIG. 34.—1, nuclei from the epidermis of the ovary-wall of *Campanula trachelium*. 2, nucleus from the spongy parenchyma of the leaf of *Melampyrum arvense*. 3, nucleus from the epidermis of the leaf of *Lophospermum scandens*. 4, nuclei from the palisade-parenchyma of the leaf of *Candollea aduata*. 5, nuclei from the wall of an almost ripe fruit of *Alectorolophus major*. 6, nuclei from the ovary of *Mimulus Tillingi*. *n*, nucleolus. *p*, protein crystalloids.

These leave any doubt only as to whether all those bodies which correspond with the undoubted crystalloids are really to be regarded as identical with them in substance; but it may be considered as certain that these are not identical with any other *known* inclusions of the nucleus.

For the recognition of crystalloids one may best fix the material with a concentrated alcoholic solution of corrosive sublimate (§ 310) and stain with acid fuchsin, or use a double staining of acid fuchsin and hæmatoxylin. As to the staining with acid fuchsin, which is also known as "Fuchsin S after Weigert," this may be conducted in various ways; but the three following methods have proved best heretofore (cf. Zimmermann II, 12 and 55, and III, 113).

a. Altmann's Acid Fuchsin Staining.

345. This is chiefly useful for microtome sections. These are fastened to the slide and, after the removal of the paraffine, are covered with a solution of 20 grams of acid fuchsin in 100 ccm. of aniline-water,* and then warmed until the

---

* This solution keeps well and only needs to be filtered occasionally.

under side of the slide is very hot to the touch. Boiling the solution is to be avoided, though its complete drying up does not injure the staining. When the solution has acted from two to five minutes, or longer, it is rinsed off with a mixture of one part alcoholic picric acid solution and two parts water, and the washing with this solution is, in general, continued until the sections no longer give off any visible color. But in many cases various degrees of staining may be obtained by stopping the washing earlier or later. The picric acid is finally removed with alcohol, and the preparation mounted in Canada balsam in the usual way.

It may be observed that, according to Altmann's directions, the sections should be gently warmed again in picric acid solution after being washed with it, in order to obtain good differentiation. But with vegetable objects I have in most cases obtained no good results from this warming with picric acid, while otherwise this method has repeatedly done me good service.

With this method a very deep staining of the nuclear crystalloids is always obtained ; but it is inferior to the two following methods in that, especially in young cells, at least when fixed with sublimate, the nucleolus is also pretty deeply stained.

b. Acid Fuchsin Method B.

346. The second method, which I have called "acid fuchsin method B" (cf. Zimmermann II, 14), is adapted as well for free-hand sections as for those from the microtome. The well-washed sections are placed first in a .2% solution of acid fuchsin in distilled water, to which a little camphor is added to make it keep better. They remain in this solution at least several hours, best 24 hours or longer.* They are then washed as quickly as possible in running water (cf. § 39). The time necessary for this differs much in different cases, and varies between a few minutes and several hours. But it

---

* If many sections are to be stained at the same time in this way, the vessel described in § 37 may be used.

may be determined easily by a couple of trials. After washing, the preparations are transferred in the usual way to Canada balsam.

By the use of this method I have obtained a very good staining of the nuclear crystalloids in both free hand and microtome sections. They are always deeply stained, even if all the other constituents of the nuclei, even the nucleoli, have been decolorized long before.

c. The Acid-fuchsin-Potassium-bichromate Method.

347. The third method, which may perhaps be briefly called the "acid fuchsin method C," agrees almost completely with Altmann's method (§ 345) in that here also the microtome sections are warmed with a concentrated solution of acid fuchsin. But for washing a warmed solution of potassium bichromate is used instead of picric acid, and the result depends neither on the concentration of the solution nor on the maintenance of a definite temperature. I used mostly a concentrated aqueous solution of the salt heated in the paraffine oven to 50° or 60° C., but even a boiling solution may be used. When the preparations are sufficiently washed, which can usually be readily recognized after a little practice, but may be determined by a few trials, the potassium bichromate is quickly washed with water and then the preparation is transferred to balsam in the usual way.

By the use of this method I obtained always very clear staining of the nuclear crystalloids. These remained still deeply stained when the color had long been washed out of the nucleolus.

d. Double Staining with Acid Fuchsin and Hæmatoxylin.

348. This method may be used in very different ways. But I have generally found it best to first stain the objects in mass with Delafield's hæmatoxylin (§ 314), and then to stain microtome sections cut from them with acid fuchsin by the method B (§ 346). Then, in one and the same nucleus may be seen the deep red-colored crystalloids beside the deep blue-violet nucleolus, within the violet nuclear frame-

work. This method has also proven good for following the fate of the crystalloids during karyokinesis (Zimmermann, III, 119).

2. The Centrospheres.

348a. After the presence of the so-called centrospheres * or attractive spheres had been recognized in animal cells by various authors (cf. Flemming I and III), Guignard (IV) succeeded in observing them in various plant-cells also, and it is now to be regarded as not improbable that these bodies are constant constituents of the cell.

The centrospheres, which have also been called by Guignard "*sphères directrices*," consist of a generally globular central portion (the "centrosome" of Guignard) which is surrounded by an unstainable envelope, and occur usually

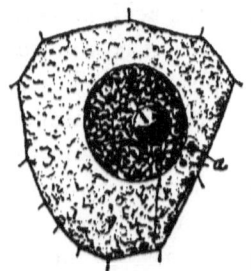

FIG. 35.—Embryo-sac of *Lilium Martagon* before the first nuclear division. *a*, centrospheres. After Guignard.

a pair in each cell (cf. Fig. 35 and 36, *a*). They appear to play an important part, especially in karyokinesis. At least, they form, after their separation, the centres of the radiating structures observed in the cytoplasm (cf. Fig. 36, II); and the threads of the achromatic spindle (cf. § 342) run together at both ends at the attractive spheres (cf. Fig. 36, III and IV). At the same time with the splitting of the segments of the nuclear threads, a division of the centrospheres occurs, so that each daughter-nucleus has again two attractive spheres (cf. Fig. 36, IV), which at first remain together, and separate only at the beginning of another nuclear division.

In the sexual act of the Angiosperms, according to Guignard's observations, the attractive spheres of the male nucleus enter the egg-cell at the same time with it, and the spheres of male and female origin fuse in pairs (cf. Fig. 37).

---

*[This English equivalent of the term lately proposed by Strasburger for these structures seems, on the whole, the most available name for them. The non-staining envelope of the centrosome may be termed, with Strasburger, the *astrosphere*.]

SPECIAL METHODS.

348b. Guignard recommends (IV, 166), for bringing out the centrospheres, *fixing* with alcohol, a 20–30% alcoholic solution of corrosive sublimate or picric acid, a 1% aqueous solution of corrosive sublimate, a saturated aqueous solution of picric acid, or a .5% solution of chromic acid. He has

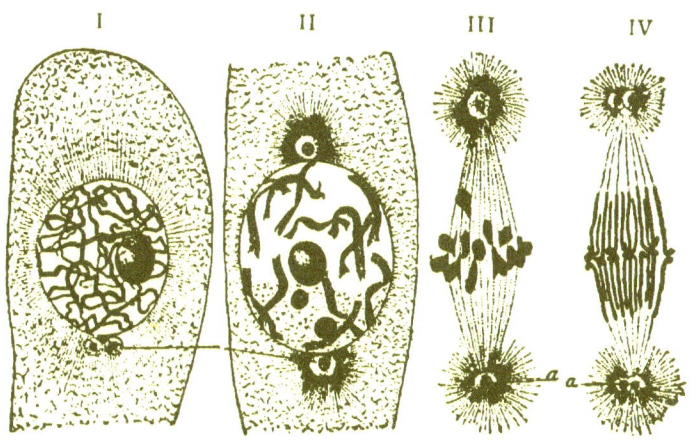

FIG. 36.—*Lilium Martagon*. I, tip of the embryo-sac; II, the same, later stage; III and IV, older karyokinetic figures from the same source; *a*, centrospheres. After Guignard.

also used the vapor of osmic acid, but allows it to act only a short time, in order not to lessen the staining capacity of the objects, and then places them in Flemming's solution (§ 309) for half an hour to an hour, and then in alcohol.

For *staining* the attractive spheres Guignard uses especially hæmatoxylin; but he first treats the sections hardened with alcohol with a 10% solution of *zinc sulphate* or ammonia alum. He has also treated the preparations successively with a dilute aqueous solution of orseillin and eosin-hæmatoxylin.* The

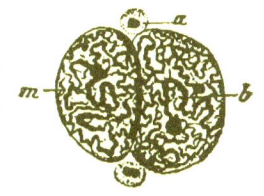

FIG. 37.—Nuclei from the fertilized egg-cell of *Lilium Martagon* during their fusion. *m*, male, *b*, female nucleus; *a*, centrospheres. After Guignard.

---

* This probably means the eosin-hæmatoxylin mixture recommended by Renault. It is prepared, according to Gierke (I, 86), by mixing equal parts of glycerine, containing common salt and saturated with eosin, and a saturated solution of potash alum in glycerine. This mixture is filtered and then an alcoholic solution of hæmatoxylin or Delafield's hæmatoxylin (§ 314) is added.

interior of the centrosphere is stained deep red by this treatment.

For *mounting* such preparations, Guignard recommends especially glycerine-gelatine and a 10% solution of chloral hydrate thickened with gelatine. The latter has the advantage of clearing the preparations, but gradually destroys most dyes.

348c. Guignard has succeeded in bringing out the attractive spheres especially in various sexual cells. In the growing stamen-hairs of *Tradescantia* he also succeeded by treating them successively with osmic acid vapor, Flemming's chrom-osmic-acetic acid mixture, and alcohol, and then staining with a mixture of fuchsin and methyl green. If this mixture is rightly prepared, the centrospheres are colored bright red in the pale red protoplasm.

348d. Hermann (II, 583) has recently used the following method for making visible the centrospheres and the radiating structures around them, in animal cells. The objects, fixed with platinum-chloride-osmic-acetic acid and then reduced with wood-spirit in the manner described in § 313, are placed whole in the dark in a hæmatoxylin solution containing one part hæmatoxylin, 70 parts alcohol, and 30 parts water. They remain in this solution 12 to 18 hours, are then treated for the same time, also in the dark, with 70% alcohol, and are then imbedded and sectioned with the microtome. The sections are then extracted with a solution of *potassium permanganate* so dilute that it has a bright rose color, until they have an ochre-colored appearance. After rapid rinsing in water, the manganese peroxide is dissolved out with a solution of one part oxalic acid and one part potassium sulphate to 1000 to 2000 parts of water, and the sections are then stained for three to five minutes with safranin. The attractive spheres and the structures surrounding them appear deeply blackened, while the nuclear elements have a bright red color.

How far this method can be used with success for plant-cells remains to be shown. But I will remark that the methods used by Flemming on animal cells with the best results (cf. § 323) are poorly suited to plant-cells, according to Guignard (IV, 167). I have obtained, also, in some not very extended experiments with Hermann's methods, no staining of the centrospheres, while the spindle-threads of such preparations stood out very sharply, especially after staining with gentian violet.

3. The Chromatophores and their Inclusions.

349. Under the name chromatophores are commonly included at present three different kinds of bodies; the green chlorophyll-bodies, *chloroplasts*, and the corresponding bodies in the algæ which are not green, the mostly yellow or red bodies which carry coloring matters, *chromoplasts*, occurring especially in the bright-colored parts of flowers and fruits, and the colorless *leucoplasts*, which are found chiefly in subterranean and young parts of plants. The grouping of these different bodies together is justified, aside from their chemical similarity, by the fact, recognized especially by Schimper, that they stand in genetic relations with each other, and may pass over into each other in the most various ways

I. *Methods of Investigation.*

350. The study of chromatophores has been conducted chiefly in the living cell. Of course this is only possible in sections which are at least several cell-layers in thickness; and the most rapid preparation possible is necessary, since chromatophores are very sensitive to the most varied harmful influences. Since most cells also die very quickly in pure water, and the chromatophores especially suffer profound structural changes in this medium, it is advantageous to use a dilute solution of salt or sugar as a medium for their study. I have used with good results a 5% solution of sugar, with which I injected the tissues to remove the air from the intercellular spaces, which may usually be easily done by means of a filter pump (cf. also § 5).

351. In difficult cases one must have recourse to staining methods. I have found a concentrated alcoholic solution of

corrosive sublimate well adapted for fixing (cf. § 310); and a concentrated alcoholic picric acid solution often does well.

According to my own most recent experiments, a saturated solution of picric acid and corrosive sublimate in absolute alcohol seems to be best for fixing chromatophores. I allow it to act about 24 hours on the objects to be fixed and wash it out with running water. The use of an iodine solution for the removal of the sublimate seems unnecessary here, as I have seen none of the well-known sublimate needles in my preparations.

Krasser (II, 4) recommends the use of a 1% alcoholic solution of *salicylic aldehyde* for fixing chromatophores. He lets it act for 24 to 48 hours on small pieces of tissue. After hardening in alcohol, the sections may be mounted in glycerine, glycerine-gelatine, or balsam. If in the latter, the clearing in clove-oil must be made as brief as possible.

352. Schimper used hæmatoxylin and gentian violet for staining chromatophores; but I have found iodine green, fuchsin, and acid fuchsin better (cf. Zimmermann V, 6).

Staining with *acid fuchsin* is best accomplished by one of the three methods described in §§ 345 to 347. It is easy to make clearly visible the relatively small leucoplasts on each starch-grain in the outer layers of a ripe potato, by the aid of method B (cf. Fig. 38, *l*).

353. *Iodine green* is used in concentrated aqueous solution and is either allowed to act for only a short time ($\frac{1}{2}$ to a few minutes) on microtome sections, which are then washed with water and examined in glycerine; or it is allowed to act longer and the sections are then placed in a solution prepared by mixing two parts of common ammonia with 98 parts of water. In this the sections were left

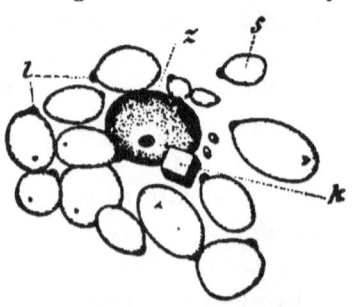

FIG. 38.—Cell-contents from a parenchyma-cell of a tuber of *Solanum tuberosum*, but a few layers removed from the cork. After fixing with sublimate-alcohol and staining with acid fuchsin (Method B). *l*, leucoplasts; *s*, starch-grains; *z*, nucleus; *k*, crystalloid.

from a few minutes to several hours, according to the depth of the staining and the character of the preparation. The examination may be made in glycerine. Such preparations keep only a very short time, while very permanent preparations may be made by transferring to Canada balsam even sections stained with iodine green. In this transfer alcohol must be wholly avoided, since it decolorizes the chromatophores. Phenol and aniline also are not suited for this use. Therefore I simply allowed the sections to dry, after washing them with water, and then treated them with xylol, and finally added xylol-balsam.

354. For staining with *fuchsin*, I have used the ammonia fuchsin mentioned in § 271. I let it act only a short time on the sections, until they begin to become red. Then I wash it out with water, and examine the sections in glycerine or transfer them in the above described manner, by drying, to balsam. Alcohol decolorizes the chromatophores in this case, also.

II. *The Finer Structure of the Chromatophores.*

355. Opinions are at present divided as to the intimate structure of the chromatophores (cf. Zimmermann I, 56 and Bredow I, 380, for the early literature). It is only as to the

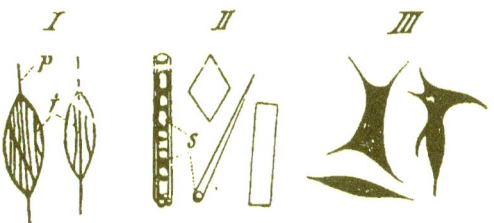

FIG. 39.—I, chromoplasts from the flower of *Neottia nidus-avis*; *p*, protein crystalloids; *f*, pigment-crystals. II, the same, from the root of *Daucus Carota*. III, the same, from the fruit of *Sorbus aucuparia*. *s*, starch-grains.—After Schimper.

chromoplasts that it may be regarded as settled that the pigment occurs partly in crystalline, partly in amorphous form.

356. In the former case, it forms more or less regular

rhombic plates or peculiar cylindrically curved bodies, as, for example, in the parenchyma of the carrot (cf. Fig. 39, II); or it occurs in the form of delicate needles which are imbedded in the colorless stroma in small numbers, as in *Neottia nidus-avis* (Fig. 39, I), or in large quantity, as in the pericarp of *Sorbus aucuparia* (Fig. 39, III). All these pigment-crystals, which consist of carotin, according to the prevailing nomenclature, are characterized by strong pleochroism *), and this peculiarity has been used by Schimper (III, 94) in difficult cases for the recognition of the crystalline structure of the fine pigment-needles.

357. The amorphous pigment occurs within strongly refractive globules ("*grana*") which are imbedded in the usually quite colorless stroma. The flesh-colored fertile stems of *Equisetum arvense* contain chromatophores with especially large grana (Fig. 40).

The study of the crystalline and amorphous pigments can, of course, be carried on with entire certainty only within the living cells, and, on account of their ready decomposition, the precautions described in § 350 should be very strictly adhered to.

F1G. 40. — Chromoplasts from the parenchyma of the stem of *Equisetum arvense*. (×1400.)

III. *The Inclusions of the Chromatophores.*

358. The most widely distributed of the inclusions of the chromatophores, the starch-grains, will be treated in connection with various related bodies in § 400 and following ones. Besides these, there have been recognized in the chromatophores protein crystalloids, leucosomes, pyrenoids, and oil-drops, which will now be discussed in order.

---

* This is determined by observing the body with one nicol prism, preferably the polarizer, and turning either the nicol or the crystal. Pleochroïc objects then appear quite colorless in one position, or at least very light colored, and, after rotation through 90°, more deeply colored than without the use of the nicol.

a. Protein Crystalloids.

359. The protein crystalloids observed in the chromatophores, especially by Schimper (III) and A. Meyer (II), often form elongated prisms (cf. Fig. 41, 2) or needles (Fig. 41, 1 and Fig. 39, I, *p*); but they are not rarely more octahedral or more or less irregular (cf. Fig. 41, 1, 3, and 4).

With the exception of those of *Canna*, they are soluble in water, but are fixed by fixing media for proteids. I have found the same methods used for the study of the nuclear crystalloids (cf. §§ 345 to 347) well suited to their investigation. Fixing with an alcoholic solution of corrosive sublimate and staining with acid fuchsin by method C has proved the best treatment. In this way it is pretty easy to obtain preparations in which the stroma of the chromatophores is quite colorless, while the crystalloids are colored deep red

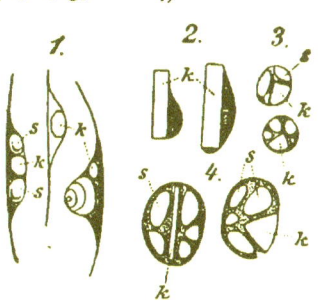

FIG. 41.—1, leucoplasts from a young shoot of *Canna Warszewiczii* with crystalline needles and octahedral crystals; 2, chloroplasts from the epidermis of the petiole of *Hedera* sp.; 3, chloroplasts from the palisade-parenchyma of the leaf of *Convolvulus tricolor*; 4, chloroplasts from the palisade-parenchyma of the leaf of *Achyranthes Verschaffelti*; *k*, protein crystalloids; *s*, starch-grains. 1, after Schimper.

b. Leucosomes.

360. Within the leucoplasts of the epidermis of various species of *Tradescantia*, and in various other plants, I have found globular inclusions which I have provisionally termed leucosomes (cf. Zimmermann II, 3 and III, 147). These consist, at least chiefly, of protein-like substance and are probably closely related chemically to the protein crystalloids already described, with which they also correspond in their behavior with acid fuchsin.

In favorable cases, as, for instance, in the epidermis of the leaves of *Tradescantia discolor*, the leucosomes may be well observed within the living cell, and tangential sections are especially adapted to this study. They occur here mostly singly or in twos or threes in each leucoplast (cf. Fig. 42, *l*).

But, since the leucosomes are soluble in water and very sensitive to the most various reagents, it is better to use, as a rule, fixed and stained material.

Fig. 42.—Leucoplasts from the epidermis of the upper surface of the leaf of *Tradescantia discolor*. *l*, leucosomes.

The best method is to fix sections from the living objects in a concentrated alcoholic solution of corrosive sublimate, and, after washing (cf. § 310), to stain by the acid fuchsin method B. If the washing is stopped at the right moment, the leucosomes alone are deeply stained.

c. Pyrenoids.

361. The starch-centres or pyrenoids observed in the chromatophores of various *Algæ* and of *Anthoceros* consist of a central portion of proteid and a starch-envelope surrounding it.

The most essential part of the pyrenoid, the proteid nucleus, appears often to possess a more or less regular crystalline form (cf. Fig. 43, 2), according to Schimper's observations (III, 74). On the other hand, it is certain that non-crystalline pyrenoids also occur, as in case of the one shown in Fig. 43, 3, which probably represents a stage in fission (cf. Zimmermann I, 48). Similar forms have lately been observed by Klebahn (I, 426) in *Cosmarium*.

362. In the presence of large quantities of starch, the

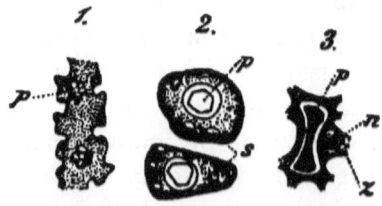

Fig. 43.—1, part of a chromatophore of *Spirogyra* sp.; 2, chromatophores of *Cladophora* sp.; 3, constricted pyrenoid of *Zygnema* sp. after fixing with picric acid and staining with acid fuchsin; *p*, pyrenoid; *s*, starch-grains; *n*, nucleus; *n*, nucleolus. 2, after Schimper.

proteïd nucleus of the pyrenoid is only with difficulty or not at all to be recognized within the living cell. Then the use of suitable staining media is to be recommended. The pyrenoids behave with these in about the same way as do the crystalloids of the nuclei and the chromatophores. Very deep staining of them may be obtained by fixing algæ with picro-sulphuric acid (§ 304) or an alcoholic solution of corrosive sublimate (§ 310), washing well, and then leaving them at least 24 hours in a .2% solution of *acid fuchsin*. After washing in water for a quarter of an hour and dehydrating in Schulze's apparatus (§ 16), the objects should be transferred to Canada balsam by the aid of Schulze's settling cylinder (§ 21). In good preparations only the pyrenoids are deep red, and even the nucleoli are quite colorless.

363. A simultaneous fixing and staining of chromatophores and pyrenoids may be obtained more simply by placing the algæ in a concentrated solution of picric acid in 50% alcohol, to which a little acid fuchsin is added (about five drops of Altmann's solution, described in § 345, to a watch-glass full of the picric acid solution). In this they remain for two hours or longer, and are then washed for a quarter of an hour in alcohol (not previously in water) and then transferred as quickly as possible to balsam by the aid of Schulze's settling cylinder (§ 21). Preparations thus obtained appear to keep well, while a gradual fading occurs in those mounted in Vosseler's turpentine (§ 27).

363a. According to the researches of Hieronymus (I, 358), the pyrenoids of *Dicranochæte reniformis* appear to have a more complicated constitution. This author states that they consist of a protein crystalloid and a proteïd-like envelope, while the starch-grains are formed quite independently of the pyrenoids, at any part of the chromatophore.

The crystalloids are soluble, according to Hieronymus, in boiling water (!), dilute and concentrated caustic potash solution, common salt solution, acetic acid, and hydrochloric acid. After previous treatment with alcohol, its solubility becomes less. After lying for several days in a mixture of one volume of pepsin-glycerine and three volumes of .2% hydrochloric acid, the crystalloids become very transparent. Hieronymus recommends *safranin* for staining them.

The envelope is also soluble in concentrated and dilute

caustic potash solution, as well as in hydrochloric acid of various strengths. But it is not attacked by the above-mentioned digesting fluid, even after lying in it for days. *Hæmatoxylin* and Congo red are specially adapted to staining the envelope. By the combination of safranin and hæmatoxylin, Hieronymus obtained preparations in which the envelope was stained violet and the crystalloid deep red.

d. Oil-drops.

364. Drops of an oil-like substance are pretty widely distributed within chromatophores, as has been shown especially by A. Meyer (II), and Schimper (III, 106). According to A. Meyer's researches (II. 17), however, these do not fully correspond in their chemical relations either with the true fatty oils or with the ethereal oils. It is especially noteworthy that they are soluble even in dilute alcohol, but insoluble in acetic acid. They are, further, only very gradually browned by osmic acid, and are always insoluble in water, but readily soluble in ether. They behave variously with chloral hydrate. According to all these reactions, the bodies in question must surely be nearly related to the fatty oils, so much the more that they stain deeply with alcannin and cyanin by the use of the methods previously described (§§ 109, 110). At all events, it seems to me best to designate these bodies as oil-drops, as has usually been done in the literature, so long as no exact investigations of them exist.

I can recommend as a suitable object for the study of the reactions of these oil-drops, the epidermis of the leaf of *Agave americana*. This contains leucoplasts which always enclose pretty large oil-drops, especially in old leaves (cf. Fig. 44). In yellowed leaves these have usually run together into a single large drop, beside which the rest of the leucoplast is so inconspicuous that it cannot usually be directly made out with certainty, by direct observation.

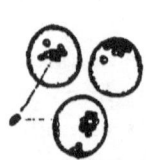

FIG. 44.—Leucoplasts from the epidermis of an adult leaf of *Agave americana*. *o*, oil-drops.

The oil-drops observed in the chromatophores of older

leaves especially are usually pretty deep yellow. This is due to a pigment belonging to the group of lipochromes, which is perhaps identical with xanthin (cf. Zimmermann III 102).

4. The Eye-spot.

365. The name eye-spot or stigma is given to the red or brownish body observed in various motile Algæ and swarm-spores, which usually occur singly at the forward end of the organism. These bodies correspond with the chromatophores in possessing a protoplasmic stroma filled with pigment. This pigment appears, according to the investigations which have been made, to be identical with carotin (§ 170) in the green algæ which have an eye-spot; at least, it shows, according to Klebs (I, 30), the characteristic blue color with sulphuric acid, in *Euglena*.

The study of the eye-spot has heretofore been carried on almost exclusively on living material.

5. The Elaioplasts and Oil-bodies.

366. Wakker gives (I, 475) the name elaioplasts or oil-formers to bodies contained in the protoplasts, which he has observed in the epidermis of young leaves and in the superficial parts of young stems and roots of *Vanilla planifolia* (cf. Fig. 45, *e*). They consist of a protoplasmic stroma and imbedded oil. The latter is deeply stained by alcannin (§ 104) or cyanin (§ 110), and escapes in drops from the elaioplasts when they are treated with an aqueous solution of picric acid, with glacial acetic acid, or with concentrated sulphuric acid; it is soluble in alcohol and in caustic potash solution.

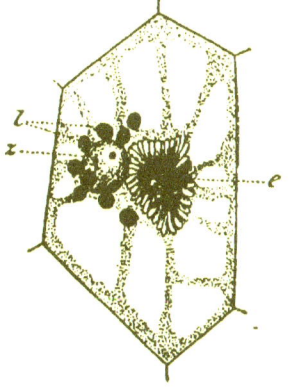

FIG. 45.—Epidermal cell of a very young leaf of *Vanilla planifolia*. *e*, elaioplast; *l*, leucoplasts; *z*, nucleus. After Wakker.

I have lately recognized elaioplasts in various other plant-tissues, especially in the epidermis of different parts of the flower of *Funkia, Ornithogalum, Agave, Dracæna,* and others (cf. Zimmermann VI, 185).

367. For *fixing* the elaioplasts Wakker recommends especially a concentrated aqueous solution of picric acid. He obtained a beautiful double *staining* of sections fixed with this fluid by using aniline blue and alcannin in the following way. He added alcanna tincture in drops to a dark blue solution of aniline blue in water, until the fluid had taken a dark purple-red color, like that of hæmatoxylin. He left the sections 20 hours in this mixture and then examined them in glycerine. The protoplasm was then light blue, the nucleus and chromatophores dark blue, the oil a fine red, and the elaioplast dark purple. These colors keep for a long time in neutral glycerine-gelatine.

368. Wakker (I, 482) includes among the elaioplasts also the oil-bodies of the *Hepaticæ,* first described by Pfeffer (VI). He was able to show by abnormal plasmolysis with a 20% solution of saltpeter (cf. §§ 432 and 436) that these always lie in the protoplasm. To render the protoplasmic wall of the oil-bodies visible, Wakker recommends, besides the dilute alcohol used for the purpose by Pfeffer, a concentrated aqueous solution of picric acid. In this case the access of the acid must be followed with the microscope, since a flattening of the protoplasmic wall of the oil-body usually results from its long action.

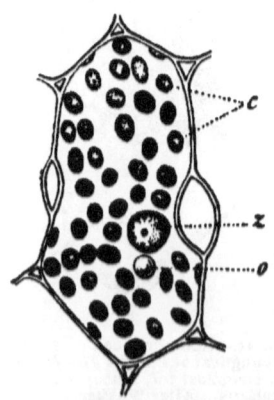

Fig. 46. — Mesophyll-cell from the leaf of *Avena orientalis.* c, chloroplasts; z, nucleus; o, oil-drops.

369. The fat-bodies and oil-bodies described by Radlkofer (I and II), Monteverde (I), and Solereder (I) are perhaps related to the oil-bodies of the Liverworts. They have been observed by these authors in the palisade and spongy parenchyma of various members of the *Cordiaceæ, Combretaceæ, Cinchoneæ, Sapotaceæ, Sapindaceæ, Gramineæ, Gaertneraceæ,* and *Rubiaceæ.* They usually occur singly in each cell, and have always about the same size in

the same individual. In the mesophyll of *Avena orientalis* they are of about the same size as the chloroplasts (cf. Fig. 46, *o*).

According to Monteverde, they lie in the protoplasm and are mostly isotropic, but in many plants are also doubly refractive, especially in dried tissues. In the grasses this double refraction disappears, according to Monteverde, when sections are warmed to 50°–55° C. in water, but reappears after a few minutes.

These oil-bodies generally behave with reagents like true fats. I have found them, especially in *Oplismenus imbecillus* and *Elymus giganteus*, insoluble in cold or hot water or alcohol, but soluble in ether, petroleum-ether, chloroform, xylol, and clove-oil. They are colored brown or black by osmic acid. They are deeply stained by cyanin (§ 110) and alcannin (§ 109). On the other hand, they remain completely unchanged in form after 24 hours in a mixture of one volume of concentrated caustic potash solution and one volume of ammonia solution, and seem not to be capable of saponification in this way (cf. § 112).

370. Besides these, Monteverde found in grasses containing crystals and in those free from them, drops of an oil-like appearance, but of wholly unknown composition. They increase in size in water, glycerine, and dilute acids only by the formation of vacuoles, but gradually dissolve in alcohol. After a long stay in water, they become insoluble also in alcohol. They dissolve with swelling in strong mineral acids and acetic acid, and instantly in caustic potash, ammonia, ether, chloroform, and chloral hydrate. They are not stained by alcanna tincture, but easily so by aniline dyes. They become brown with iodine. Monteverde considers it probable that they consist chiefly of resin.

I will mention here the structures observed by Lundström (I) in the epidermal cells of various *Potamogeton* species, which he regards as oil-drops, although, according to his statements, they are soluble in very dilute alcohol.

6. The Iridescent Plates of various Marine Algæ.

371. As was first described in detail by Berthold (II, 685), there occur in the superficial cells of some marine algæ characteristic iridescent protoplasmic plates, which are very probably to be considered as organs for protection from too strong illumination. In the *Chylocladiæ* they consist of a strongly refractive mass in which small granules of somewhat varying size are imbedded (cf. Fig. 47, *a*, *p*). In profile view

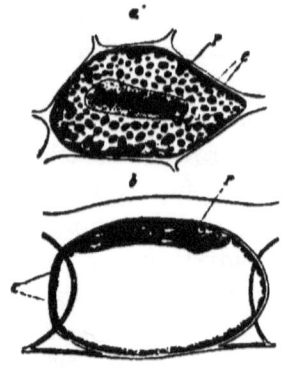

they show a striation parallel to the surface of the plates and suggesting their composition of separate lamellæ (cf. Fig. 47, *b*). These plates are always very sharply differentiated from the cytoplasm.

In distilled water vacuoles occur in them, most probably in consequence of the swelling of the globules imbedded in its mass; and finally, the plates show a completely spongy structure.

FIG. 47.— Superficial cell of the thallus of *Chylocladia reflexa*, *a*, in surface view; *b*, in profile view. *p*, iridescent plate; *c*, chromatophores (× 250). After Berthold.

372. For *fixing* the protoplasmic plates Berthold recommends chiefly a concentrated solution of iodine in sea-water (cf. § 301) or osmic acid (§ 308). Neither of these, however, completely preserves the structure.

Whether these bodies are to be included with elaioplasts, as has lately been held to be probable by Wakker (I, 488), must be determined by further researches.

7. Microsomes and Granula.

373. Under the term microsomes are usually included all those small and mostly globular bodies which are distinguishable by their different refractive power from the main mass of the cytoplasm. But it can no longer be doubted that these include bodies of very different composition, and it is not possible to speak of special reactions of the microsomes.

On the other hand, it has been shown by Altmann (I) that bodies of definite reactions may be quite generally recognized in the cytoplasm of animal cells, which this author terms *granula* and regards as the elementary organisms of the cell. Altmann used, for the demonstration of these granula, chiefly a fixing mixture of osmic acid and potassium bichromate and the acid fuchsin staining method A, described in § 345.

374. How far the cytoplasm of plant-cells possesses a similar granula-structure cannot at present be said. The writer's investigations on this point have not yet reached any conclusive results. But, by the aid of Altmann's methods, it can be shown that certain bodies are widely distributed in the cytoplasm of the cells of the assimilating tissue, which correspond in many respects with Altmann's granula and have been termed at first simply granula (cf. Zimmermann, II, 38).

These are always colorless and are mostly little spheres, which have, at most, about the size of the nucleoli, in adult cells (cf. Fig. 48, *g*). Their chemical relations indicate that they consist of protein-like substances.

375. For *fixing* the granula a concentrated alcoholic solution of corrosive sublimate or of picric acid may be used (cf. §§ 310 and 303). Very good results have been obtained also with dilute nitric acid, and I used in this case a solution containing, in 97 volumes of water, three volumes of chemically pure nitric acid of specific gravity 1.3, which therefore contained about 1.5% of HNO_3. I allowed this solution to act for 24 hours, and then washed the objects in running water for 24 hours.

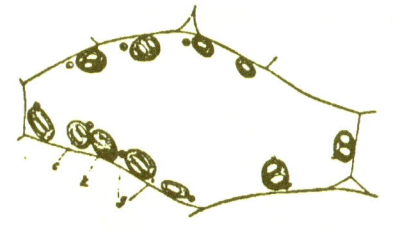

FIG. 48.—A cell from the lowest mesophyll layer of *Tradescantia albiflora*, from a preparation fixed with alcoholic picric acid and stained by Altmann's acid fuchsin method. *c*, chloroplasts; *k*, nucleus; *g*, granula.

For *staining* the granula I formerly used almost exclusively Altmann's acid fuchsin method (§ 345), but have lately convinced myself that the other acid fuchsin methods (§§ 346 and 347) give very good results. Especially after fixing with nitric acid, preparations may be pretty easily obtained in which the granula are still deeply colored, while the much larger chromatophores are wholly decolorized. I can recommend the leaves of *Tradescantia albiflora* as suitable objects for study, as they contain comparatively large granula, especially in the spongy parenchyma (cf. Fig. 48, g).

[Crato (I) has lately observed, in *Chætopteris* and other plants, certain structures, hitherto included under the general term microsomes, which he regards as special organs of the cell and calls *physodes*. For further details, his account of them may be consulted.]

8. The Cilia.

376. The cilia, which occur on most of the freely motile lower organisms and are always directly connected with the protoplasm, are often so fine that during their active motion they can be recognized with difficulty or not at all, even with the best objectives.

377. In many cases the cilia may be made better visible by bringing the organisms to rest by quickly *killing* them. For this purpose, the vapor of *osmic acid* or 1% osmic acid, 1% chromic acid, or the solution of iodine and potassium iodide may be used. The cilia often appear sharply, also, if a drop containing the organisms be allowed to dry upon the slide.

378. If the position of the cilia is to be determined while in motion, fine granules of carmine, or the like, may be added to the fluid containing the organisms, according to the method proposed by Bütschli (I, 7). The movements of these granules will show the cilia-bearing end.

379. Recently *staining* methods have also been used for the recognition of cilia.

Migula (I, 76) obtained a fine staining of the cilia of *Gonium pectorale* by using the following method: A very small drop of a concentrated alcoholic solution of cyanin was added to the living specimens, and, after a time, enough water was added to precipitate the cyanin not taken up by the organisms, in granular form. The cilia, as well as the rest of the protoplasm, are colored at first pale blue, but after the addition of water, deep violet. These methods have given me no favorable results, when used on various algæ.

380. But I have obtained a very deep staining of the cilia in *Chlamydomonas*, *Pandorina*, and *Chromophytum* by the following method, which is essentially similar to methods used for staining the cilia of *Bacteria* (cf. § 476). The objects were first fixed in the hanging drop on the slide by the fumes of osmic acid (cf. § 308), and then allowed to dry; then a drop of a 20% aqueous tannin solution is added, and washed off with water in five minutes or later. The slide is then plunged in a concentrated aqueous solution of fuchsin,* in which it remains a quarter of an hour or longer. The fuchsin solution is now washed off with water, the preparation is again allowed to dry, and finally a drop of balsam and a cover-glass are placed upon it. I have obtained in this way very beautiful permanent preparations in which the cilia were stained bright red.

[I have obtained very satisfactory stainings of the cilia of the zoöspores of various algæ and fungi by adding to the water containing them a drop or two of a 1% solution of osmic acid, and then the same amount of a strong solution in alcohol of equal parts fuchsin and methyl violet. This stains the cilia deep red almost at once; and the osmic acid need not be removed before adding the stain.]

9. Protein Grains.

381. The investigation of the aleurone or protein grains which occur in the seeds of all the higher plants is best conducted, in oily seeds, after the removal of the oil, which is often a great hindrance to their study. This may be ac-

---

* Carbol-fuchsin (§ 468) is especially useful, as it stains deeply in a few minutes.

complished in many cases, as in *Ricinus*, by placing the sections to be studied, for five minutes or longer, in absolute alcohol. Less soluble fats may be extracted with ether or with a mixture of equal parts ether and alcohol. Further methods of study will depend upon the constituent of the protein grain which it is chiefly desired to make visible; for there may be distinguished in them a proteid fundamental mass and various inclusions, protein crystalloids, globoids, and crystals of calcium oxalate. These separate constituents, which are not found simultaneously in all protein grains, but yet are widely distributed, will be discussed in order.

[But it may be well to give first two general methods recommended by Krasser (III) for making permanent preparations to show the grains and their inclusions in general.

1. The sections are fixed with an alcoholic solution of picric acid, rinsed with alcohol, stained in an alcoholic eosin solution, "toned" with alcohol, cleared with clove-oil, and mounted in balsam. The fundamental mass is stained dark red, the crystalloid is yellow, and the globoid is colorless or reddish.

2. The sections are simultaneously fixed and stained in a concentrated solution of nigrosin in an alcoholic picric-acid solution. When the fundamental substance appears blue, which may be determined by examining a section in absolute alcohol with the microscope, the sections are washed in alcohol, cleared quickly in clove-oil, and mounted in balsam. This treatment stains the fundamental mass blue, the crystalloid yellowish green, and leaves the globoid colorless.]

a. The Fundamental Mass.

382. The fundamental mass of the protein grains, which sometimes forms the bulk of the whole grain and sometimes only a thin envelope about the various inclusions, consists of proteid substances and is well suited for the study of the reactions of protein, detailed in §§ 224 to 234.

It is also always readily soluble in dilute *caustic potash* or ammonia solution and in *sodium phosphate*. The last-named reagent in concentrated aqueous solution is especially recommended by Lüdtke (I, 73).

The fundamental mass of the protein granules behaves very differently in different plants. In many, as in *Pæonia*, it is soluble in water; in others it is insoluble in it. It shows similar relations with a 10% solution of common salt and a 1% sodium carbonate solution, differing with the species of plant. The protein grains which are soluble in water are best examined in alcohol or glycerine; and, by the gradual addition of water, their solution may be observed under the microscope.

383. But the protein grains may be made insoluble by fixing media, for instance by an alcoholic solution of corrosive sublimate or of picric acid. Objects fixed in the latter fluid may be directly preserved in balsam. But it is also easy to stain the protein grains after washing out the fixing fluid, and for this purpose an aqueous solution of eosin is very useful.

384. The fundamental mass of the protein grains is bounded externally, as well as against the inclusions, by a delicate *pellicle* which is distinguished from the remaining substance of the protein grain by its insolubility in dilute alkalies and acids, but, as has been shown by Pfeffer (I, 449), also consists of albuminoid materials. According to Pfeffer, it may be well observed by gradually dissolving the fundamental mass or the inclusions by the addition of very dilute caustic potash, acetic acid, or hydrochloric acid. Lüdtke has lately recommended lime-water for the same purpose, as it first dissolves the fundamental mass of the grain, while the membrane becomes sharply visible and then dissolves after a preliminary swelling.

b. The Protein Crystalloids.

385. The crystalloids observed in the protein grains of many seeds consist always, like the fundamental mass, of

proteïds, as may easily be shown by the aid of the reactions described in §§ 224 to 234. When examined in alcohol or glycerine, they are usually hardly or not at all distinguishable from the fundamental mass, since they have about the same refractive index as it (Fig. 49, II, *a*). But after being placed in water, in which the crystalloids are always insoluble, they show clearly in consequence of their greater density (Fig. 49, II, *b*). To distinguish them from the globoids and calcium oxalate crystals, one may make use of their ready solubility in very dilute caustic potash and their power of becoming yellow or brown in a solution of iodine and potassium iodide, according to its strength.

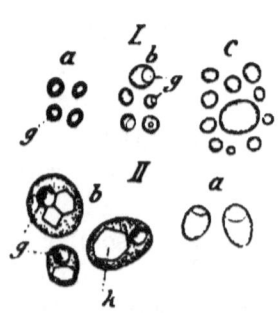

FIG. 49.—I. Protein grains of *Pæonia*, *a*, from the outer, *b*, from the middle, *c*, from the inner layers. Drawn from picric acid material.
II. Protein grains from the endosperm of *Ricinus communis*, *a*, in alcohol, *b*, on the addition of iodine-potassium-iodide solution after treatment with alcohol. *k*, crystalloid; *g*, globoid.

Lüdtke (I, 77) has lately recommended a concentrated aqueous solution of *sodium phosphate* for the recognition of crystalloids, since they are insoluble in it, while all other constituents of the protein grain are dissolved by it, though sometimes only after several hours.

The crystalloids with distinct faces belong, according to Schimper (I), partly to the isometric and partly to the hexagonal crystal-system (cf. Figs. 49, II, *b* and 50, II and III); and those of the latter system have a feeble doubly refractive power.

386. Eosin is very well adapted for staining the fixed crystalloids. Acid fuchsin may also be used for the same purpose according to one of the methods given in §§ 345-347. These dyes give a very pure and deep staining of the crystalloids, especially after fixing with corrosive sublimate.

387. Recently Overton (II, 5) and Poulsen (II) have given methods for staining crystalloids. Overton places sections of the endosperm of *Ricinus*, hardened with alcohol, first in

a dilute aqueous solution of tannin for ten minutes, and then, after careful washing, in 2% osmic acid. The crystalloids are stained a beautiful brown by these reagents. After washing out the osmic acid, the preparations may be preserved in glycerine.

Poulsen (II, 548) places the sections first in alcohol for 24 hours, then for an hour in a 25% aqueous solution of tannin, and finally, after washing this out with water, in an aqueous solution of potassium bichromate, in which he leaves them until they are brown or yellowish. For the preservation of these preparations, in which the aleurone grains should be quite transparent, Poulsen recommends glycerine.

According to another method also recommended by Poulsen, the sections, treated in the same way with alcohol and tannin, are placed, after washing, for an hour in a 10–20% aqueous solution of ferrous sulphate. The preparations are then washed and transferred to balsam in the usual way. The crystalloids then appear deep blue, almost black.

c. The Globoids.

388. The globoids consist, according to Pfeffer's researches (I, 472), of the calcium and magnesium salt of an organically combined phosphoric acid. They do not occur in all protein grains but, according to Pfeffer's investigations, are not wholly absent from any seed. They are sometimes more or less precisely globular in form, as in *Pæonia* and *Ricinus* (cf. Fig. 49, I and II, *g*), sometimes irregular, biscuit-shaped, or clustered, as in *Bertholletia excelsa* (Fig. 50, I). The relative and absolute size may vary very greatly in the same seed. For example, the protein grains in the innermost layers of the endosperm of *Pæonia* are quite free of globoids, while their size increases regularly toward the outside (cf. Fig. 49, I, *a–c*).

389. In oil or Canada balsam the globoids have the

appearance of vacuoles (Fig. 49, II, *a*), because they have a lower refractive index than these media. They may be best observed by dissolving the fundamental mass of the protein-grain and the crystalloids contained in it with dilute, about 1%, caustic potash solution, from sections previously deprived of their fat by alcohol or ether-alcohol. There remain then in the space formerly occupied by the protein-grain only the globoids and any calcium oxalate crystals that may be present. To distinguish between these two constituents, polarized light may be used. The globoids are amorphous and therefore isotropic, while the oxalate crystals (cf. § 392) are strongly doubly refractive.

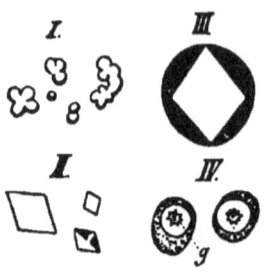

FIG. 50.—I, globoids, II, crystalloids, of *Bertholletia excelsa;* III, protein-grain of *Elæis guineensis;* IV, protein grains of *Vitis vinifera;* g, globoid with calcium oxalate crystal in the middle.

For the same purpose, a dilute, about 1%, *acetic acid* may be used, in which the crystals are insoluble, while the globoids are quickly dissolved by it. In concentrated acetic acid the globoids are soluble with much greater difficulty.

In a concentrated aqueous solution of *sodium phosphate* the globoids are completely soluble, according to Lüdtke (I, 79), even after treatment with corrosive sublimate. But this solution requires several hours, and the larger globoids, like those from the seed of *Vitis vinifera*, show during solution an evident stratification which gradually penetrates from without inwards. Lüdtke also observed similar stratifications when he allowed dilute caustic potash or lime-water to act for a long time on the globoids.

It may be remarked here that the globoids are also dissolved by picric acid. The protein-grains, however, preserve their original form completely in this fluid, and cavities may be seen in them which have exactly the forms of the dissolved globoids.

390. Pfeffer (I, 472) used the following reactions for the recognition of the chemical composition of the globoids.

The presence of organic substance in them is shown by the fact that isolated globoids blacken strongly on *heating*. They may be easily obtained by moving about on the cover-glass sections which have been freed from fats and proteids by successive treatment with ether-alcohol, 1% caustic potash, and water. To obtain a pure white ash from the globoids, very strong heating is necessary.

If the residue left after strong heating be treated with an ammoniacal solution of ammonium chloride, the characteristic crystals of ammonio-magnesium phosphate are formed. This shows at once the presence of *phosphoric acid* and *magnesium* in the globoids.

But if the globoids be treated with the ammoniacal ammonium chloride solution before being heated, no formation of ammonio-magnesium phosphate crystals occurs, evidently because the organically combined phosphoric acid behaves differently from the phosphoric acid set free by heating. But the formation of large quantities of crystals of the double salt mentioned takes place when sections, freed as above from fats and proteids, are treated with a mixture of an ammoniacal solution of ammonium chloride and sodium phosphate. I have used in this case, with good results, a reagent containing 10 parts of the two salts named and 10 parts of the officinal ammonia solution * to 100 parts of water.

391. The presence of *calcium* was shown by Pfeffer (I, 473) by the addition of an ammoniacal solution of ammonium chloride and ammonium oxalate to the unchanged globoids. There are then gradually formed the characteristic crystals of calcium oxalate. By the addition of sulphuric acid, the formation of the characteristic gypsum needles can be brought about.

d. Crystals.

392. The crystals observed within the protein-grains consists, like nearly all crystals observed within the vegetable

---

* [This is of 16° Baumé, spec. gravity .960.]

organism, of calcium oxalate. For their reactions §§ 85 to 88 may be consulted. It may be observed here that calcium oxalate crystals can occur even in the interiors of the globoids, as, for example, inside of the large protein-grains of the seed of *Vitis vinifera* (cf. Fig. 50, IV, *g*).

10. Protein Crystalloids.

393. The protein crystalloids contained in the cytoplasm or in the cell-sap agree essentially with the crystalloids contained within the nucleus, the chromatophores, and the protein-grains (cf. §§ 343, 359, and 385). Besides very regular forms, like those from the tubers of *Solanum tuberosum* (Fig. 38, *k*) or from the epidermis of the leaf of *Polypodium irreoides* (Fig. 51, I, *k*), one finds also spindle-

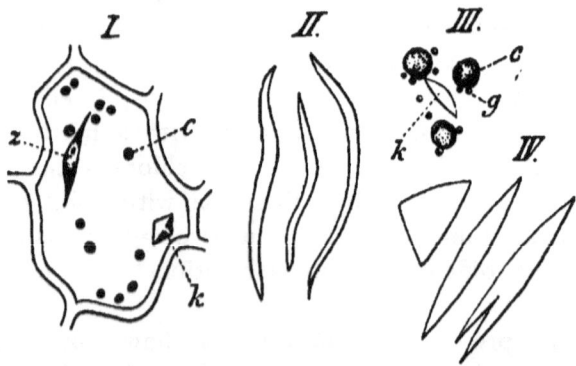

FIG. 51.—I, epidermal cell of the lower face of the leaf of *Polypodium irreoides*; *k*, crystalloid; *z*, nucleus; *c*, chloroplasts. II, crystalloids from the wall of the ovary of *Gratiola officinalis*. III, crystalloid (*k*), chloroplasts (*c*), and granula (*g*) from a spongy-parenchyma cell of *Passiflora caerulea*. IV, crystalloids from the subepidermal parenchyma of the leaf of *Vanda furva*.

shaped or needle-like forms, or such as are variously bent (Fig. 51, II–IV). Molisch (II) observed completely ring-shaped crystalloids in the leaves of various species of *Epiphyllum*.

For the recognition of these crystalloids the staining methods detailed in §§ 345 to 347 may best supplement the study of living material.

394. As has been recognized especially by J. Klein (I),

true protein crystalloids occur in many marine algæ. These must not be confused with the so-called rhodospermin-crystalloids which are formed after the algæ have been killed in a solution of common salt or in spirit.

395. *Rhodospermin* consists chiefly of hexagonal prisms or plates which are colored deep red (Fig. 52). They are insoluble in water, alcohol, glycerine, sulphuric, nitric, hydrochloric, and acetic acids, and in alkalies. They become gradually destroyed and invisible on boiling in sulphuric acid, hydrochloric acid, or potash. *Iodine* colors them at first golden yellow and later deep brown-yellow. They are not colored by concentrated nitric acid, but on the subsequent addition of ammonia become most clearly yellow. Rhodospermin often swells markedly in a *potash* solution, but contracts again to its original bulk on the addition of acids, while, at the same time, the color which has disappeared in the potash reappears (cf. J. Klein I, 55).

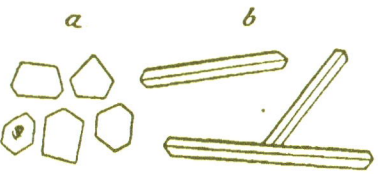

FIG. 52.—Rhodospermin crystals from *Ceramium rubrum*. *a*, formed in the cortical tissue; *b*, rhodospermin formed free in the fluid. After J. Klein.

Whether we are justified by these reactions in including rhodospermin among the proteids, as is commonly done, seems to me at least doubtful. But, at any rate, the rhodospermin crystalloids do not belong in the same category with other crystalloids, for they represent, as has already been remarked, an artificial product arising through the action of reagents.

11. Rhabdoids (Plastoids).

396. Gardiner observed (I) in most of the epidermal cells of *Drosera dichotoma*, as well as in those of *Dionæa*, spindle-shaped or needle-shaped bodies which he first termed plastoids and later, rhabdoids.* These usually occur singly in the cells, which they cross diagonally. They are fixed by

---

* From ἡ ῥάβδος, the rod.

alcohol, chromic acid, and picric acid. After fixing with the last-named acid, they may be deeply stained with Hofmann's blue; but in dilute alcohol they swell and finally disappear wholly. They are also disorganized by *iodine* and take a spherical form.

After stimulation of the leaves the rhabdoids contract and become rounded or fall into several pieces, which become at first lens-shaped, but later always more spherical.

After longer stimulation the rhabdoids decrease markedly in size; they are therefore regarded by Gardiner as reserve materials.

Whether these structures should be simply included with the crystalloids cannot be determined from our present knowledge.

12. The Acanthospheres of the Characeæ.

397. The acanthospheres,* or ciliate bodies, observed in the cells of various species of *Nitella*, have already received various explanations (cf. Zimmermann I, 73). According to Overton's recent researches (II), they consist of albuminoid substances which most probably possess a crystalline structure and are partly united with tannins.

398. For recognizing the presence of *tannin* in the acanthospheres, Overton used potassium bichromate, osmic acid, and staining *intra vitam* with methylene blue (cf. §§ 201, 203, and 208). He deduced their *proteïd* nature from their behavior with iodine and potassium iodide solution, Raspail's reagent, and Hartig's potassium ferrocyanide-ferric-chloride reagent (cf. §§ 224, 227, and 230). He also stained objects fixed with an alcoholic sublimate-solution with borax-carmine and an aqueous solution of fuchsin. The crystalline structure of the acanthospheres is best shown by the use of the former stain and mounting in *balsam Tolu*.

The difficult solubility of the acanthospheres in acids and

---

* [As I am not aware that any English name has been applied to these bodies, I propose this Greek equivalent of the German name, Stachelkügeln.]

alkalies is especially remarkable. These bodies remain almost unchanged in concentrated sulphuric, hydrochloric, and nitric acids, and in glacial acetic acid, according to Overton (II, 4). Cold caustic soda also fails to attack them, but on boiling in this solution, the spiny envelope gradually disappears and the interior assumes a spongy structure.

399. It may also be remarked that Overton has found in the cells of *Nitella syncarpa*, besides these acanthospheres, vesicles as clear as water which show the same chemical relations as they; and intermediate stages between the two bodies were to be found. In the species of *Chara* that he studied, Overton found only such spineless bodies; these are very probably identical with the strongly staining bodies recognized by the writer in the cells of a species of *Chara* not definitely determined, after fixing with nitric acid and staining with acid fuchsin (cf. Zimmermann II, 51).

13. Starch-grains and Related Bodies

a. Starch.

400. The chemical composition of starch corresponds to the formula $C_6H_{10}O_5$; but it is not yet determined whether we have to do with a completely uniform compound in the starch-grains or not.

The *form* of starch-grains shows the greatest diversity in different plants. But in a given organ of the same species of plant, only slight variations are observed, except such as are due to the different stages of development of the grains. Some of the characteristic forms of starch-grains are shown in Fig. 53. Figures I, III, and V represent simple grains from *Canna*, *Lathræa*, and *Euphorbia*. The last two sorts are especially distinguished by their characteristic form. Figure IV shows a cell from the horny part of the endosperm of corn, in which the starch-grains lie almost in contact. Figures II and VI, finally, show the so-called compound grains of *Beta* and *Smilax*. The former consist of an immense number of component grains.

401. The larger starch-grains show usually a more or less distinct *stratification*, which is certainly due to varying water-content. This may be shown by examining moist and dry starch-grains in Canada balsam or the like (cf. § 297k and Zimmermann I, 87). The silvering described in § 297n may also be carried out on starch-grains, as Correns (III, 331) has shown. It is best done by drying at 50° C. starch-grains

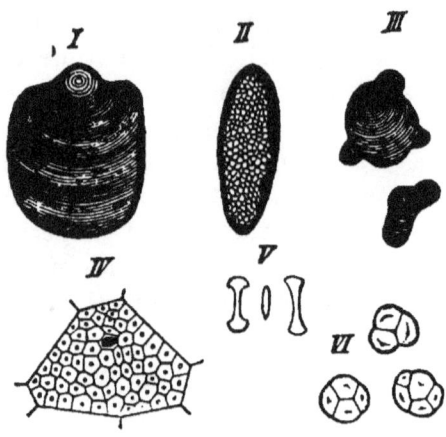

FIG. 53.—I, starch-grain from the rhizome of *Canna indica* (× 300); II, the same, from the seed of *Beta vulgaris* (× 500); III, the same, from the scale-leaves of *Lathræa squamaria* (× 150); IV, endosperm-cell of *Zea Mays*; V, starch-grains from the latex of *Euphorbia splendens* (× 150); VI, the same, from the Sarsaparilla root (× 250).

obtained by scraping and cleaned by washing and decanting. When dry, the grains are covered with a few drops of a 5% solution of silver nitrate, then a large quanity of a 10% solution of common salt is added, and the whole is exposed to bright light until the reduction is completed. The grains are finally dried again on the slide and mounted in Canada balsam. There are then seen many granules of silver deposited in the more strongly swelling layers of most of the larger grains.

402. In *polarized light* starch-grains give a characteristic figure, like that of sphærocrystals; but the orientation of the optical axes is opposite to that of the sphærites of inulin, for example. In excentrically formed starch-grains, the middle of the black cross is also excentric and always corresponds with the organic centre of the grain.

403. Concerning the *microchemical relations* of starch, it may be first remarked that it is entirely insoluble in cold water; but in hot water it suffers first strong swelling, and on longer boiling it passes completely into solution (paste). Caustic potash also causes a marked swelling and finally complete solution of the starch-grains.

But its becoming blue with *iodine* is especially characteristic of starch. This does not take place indiscriminately with iodine solutions, for all solutions containing hydriodic acid or potassium iodide give it rather violet-brown tones. A pure blue coloring of starch-grains is obtained by preparing an aqueous solution of iodine immediately before it is to be used, by adding a few drops of an alcholic solution, which keeps indefinitely, to a few ccm. of distilled water. The use of dilute solutions is always to be recommended for coloring starch-grains, as they often become almost black in concentrated solutions. Swollen grains, and even paste, have the same power as unchanged ones of becoming blue with iodine. This color disappears on heating, but reappears on subsequent cooling, without further addition of iodine. Caustic potash always at once destroys the color by the decomposition of the iodine.

404. For the recognition of very small quantities of starch various methods have been given, which consist essentially in swelling the starch-grains before the addition of iodine and destroying the rest of the cell-contents. But in most cases it is as well to place thin sections in a concentrated solution of iodine and potassium iodide; the deep black starch-grains then come out sharply with high powers and strong illuminations.

405. In many cases, the sections may well be treated according to the method of A. Meyer (II). After treatment with a pretty dilute solution of iodine and potassium iodide, and the removal of this reagent, a concentrated aqueous solution of *chloral hydrate* is added to the sections, which destroys the other cell-contents and causes the starch-grains to swell so that they appear brighter and usually of a beautiful blue color. But it should be observed that starch

is also finally decomposed by chloral hydrate, and that very small amounts of starch, which are naturally first attacked, may be easily overlooked, if the sections are not examined soon after the addition of chlorate hydrate.

The action of chloral hydrate may be hastened by heating, but the color of the starch, due to iodine, thereupon disappears, to reappear on cooling again.

Eau de Javelle (§ 12,4), recommended by Heinricher (I) for the recognition of starch, acts in the same way as chloral hydrate. But, according to this author's statements, it is better to treat the objects with it before the addition of iodine, since otherwise the iodine hinders the destruction of the protoplasm.

Both media, chloral hydrate and *eau de Javelle*, may be used with advantage when it is desired to demonstrate the distribution of starch in large organs, like whole leaves. It is best, in this case, to place the objects in a concentrated solution of chloral hydrate containing iodine, and to heat them to boiling in it. In this way leaves at all delicate may be cleared and saturated with iodine in a few minutes.

406. The recognition of starch may be accomplished with greater certainty in microtome sections. For this purpose, it is best to use a very concentrated solution of iodine and iodide of potassium, which stains the smallest starch-grains dark violet or quite black. These then stand out sharply, even in cells rich in protoplasm, from the brown proteids. In doubtful cases, a concentrated chloral hydrate solution may be afterwards added, in which the starch-grains swell and become of a beautiful blue color, while the rest of the cell-contents suffer a wide-reaching destruction.

407. I will remark here that not all starch-grains are colored blue by iodine. But there are some which are colored wholly or in part red by iodine, or which take intermediate colors between red and blue. It is probable, according to the researches of Shimoyama (I) and A. Meyer (I), that the varying behavior of these starch-grains is due to the fact that they contain greater or less quantities of amylodextrine and dextrine besides true starch.

Amylodextrine is almost insoluble in cold water, but is easily dissolved by water at 60° C., and is not thrown down from such a solution on cooling. But, on evaporation or freezing, amylodextrine separates from its solution in the form of characteristic crystalline disks, the so-called "disco-crystals" (cf. W. Naegeli I, and Naegeli and Schwendener I, 359).

408. *Amylodextrine* also arises as an intermediate product in the transformation of starch into dextrine and maltose. If starch-grains are heated for some time at 50° C. in the salivary ferment, they lose, usually after a few hours, the power to become blue with iodine, and take at first a violet, then a wine-red, and finally a pure yellow color, according to the length of time the ferment has acted. Nevertheless the grains which remain, the so-called "starch-skeletons," still have the original form of the starch-grains and often show distinct stratification. These starch-skeletons, consisting of amylodextrine, may also be obtained by the action of dilute hydrochloric or sulphuric acid for years upon unchanged starch-grains.

409. But the solution of starch within the living plant takes place in another way. As has lately been described in detail by Krabbe (I), either there occurs a regular solution from without inwards, or local corrosions are formed in the shape of pit- or crater-shaped depressions. In small grains pore-canals are often formed, which may penetrate to the interior of the grain and lead to the formation of a central cavity.

Diastase acts in a similar manner. It may be prepared by the solution of malt extract in water. A very useful ferment-containing fluid may be prepared by adding to the aqueous solution of commercial diastase about .05% of citric acid (cf. Detmer I, 197).

410. Finally there may be mentioned here a substance commonly termed *soluble* or *amorphous starch*, which agrees with true starch in being colored blue or violet to red by iodine solutions. But it is soluble in water and occurs only in the cell-sap of the epidermal cells of a few plants; for ex-

ample, *Saponaria officinalis*. No certain statements can be made as to its chemical composition (cf. J. Dufour II).

b. Floridean Starch.

411. Colorless granules have been observed in the cells of the *Floridea*, which correspond, as Van Tieghem (I, 804) recognized, in most of their chemical characters, especially in their behavior with caustic potash and hot water, with ordinary starch, and which show, under the polarizing microscope, a similarly oriented cross to that of the latter. With iodine these granules, which are commonly termed Floridean or Rhodophycean starch, usually take, however, only a yellowish-brown to a brownish-red color. According to more recent investigations of Belzung (I, 224), in many *Floridea* the younger starch-grains, especially, are colored blue by iodine. We have at present no more exact chemical studies of the substance of Floridean starch.

c. Phæophycean Starch.

412. Schmitz (I, 154 and II, 60) has observed in the cytoplasm of the *Phæophyceæ* colorless bodies quite insoluble in water, but which are not colored at all by iodine. Berthold has disputed (I, 57) the occurrence of such bodies.

d. Paramylum.

413. The paramylum-grains, recognized in the cytoplasm of the *Euglenæ* and by Zopf (I, 17) in the amœbæ and cysts of *Leptophrys vorax*, have mostly a disk-shaped or rod-shaped form (cf. Fig. 54, I); but ring-shaped grains have also been seen (Fig. 54, II and III).

Klebs (I, 271) observed an evident stratification in the large paramylum-grains of *Euglena Ehrenbergii*, which is, however, very characteristic in its arrangement and differs markedly from that of the starch-grains. These paramylum-grains consist, as is shown in Fig. 54, IV, *a–c*, which

FIG. 54.—Paramylum-grains. I, of *Euglena acus*; II, of *Phacus parvula*; III, of *Euglena Spirogyra*; IV, *a–c*, of *Euglena Ehrenbergii* seen from three sides (× 400). After Klebs.

represents such a grain in three different views, of plates placed close together and themselves composed of concentric rings. The whole grain therefore lacks a common centre of stratification. In other paramylum-grains Klebs was able to bring out the stratification only by swelling media.

The paramylum-grains differ *chemically* from the starch-grains in not being colored by iodine solutions and in not being stainable, in general. Klebs (I, 270) also gives it as characteristic of them that they remain quite unchanged in 5% caustic potash and do not swell, while in even a 6% solution of the same they at once dissolve with strong swelling. We have no more exact chemical analysis of these grains.

e. Cellulin-grains.

414. The cellulin-grains discovered by Pringsheim (I) in the hyphæ of various *Saprolegniaceæ* have sometimes the form of round or polyhedral plates and are sometimes more globular, and often show evident stratification (cf. Fig. 55). They are not colored by iodine solutions, and are even insoluble in concentrated caustic potash solution, but soluble in concentrated sulphuric acid and in a solution of *zinc chloride*. Nothing is known of their chemical constitution.

FIG. 55.—Cellulin-grains of *Leptomitus lacteus*. After Pringsheim.

Weber van Bosse (I) discovered globular bodies which agree completely with cellulose in their reactions, in one of the *Phyllosiphonaceæ, Phytophysa Treubii*. They are not colored by a solution of iodine and iodide of potassium, but are colored blue by iodine and sulphuric acid, and violet by chloroïodide of zinc.

f. Fibrosin-bodies.

415. Zopf (III) found, in the conidia of *Podosphæra Oxyacanthæ* and of other *Erysipheæ*, characteristic bodies which he calls fibrosin-bodies. They are always imbedded in the cytoplasm and are sometimes cup-shaped, sometimes in the

form of a hollow cone or a hollow cylinder (cf. Fig. 56). Their largest diameter varies from 2 to 8 μ. No stratification can be observed in them, even after treatment with caustic potash or chromic acid. For the examination of the fibrosin bodies Zopf recommends that they be isolated by pressure on the cover-glass, or that the spores be made more transparent with nitric acid or caustic potash. Zopf has determined the following points as to their chemical relations: They swell in hot water into roundish bodies; they are colored neither by iodine-potassium-iodide solution nor by chloro-iodide of zinc; neither are they dissolved by the latter reagent. In concentrated sulphuric acid they are soluble with difficulty; they are not dissolved by nitric acid, even after 48 hours' exposure. In *caustic potash* they are insoluble in the cold, but, on warming in it, they swell up into irregular, strongly refractive bodies. They are insoluble in cuprammonia, alcohol, ether, and chloroform, are not browned by osmic acid, and take up no aniline dye.

FIG. 56. — Fibrosin-bodies from the conidia of *Podosphæra oxyacanthæ*, partly in different views as indicated by the dotted lines. (× 1000). After Zopf.

The fibrosin-bodies therefore agree most closely in their reactions with fungus-cellulose, and are distinguished from the cellulin-grains especially by their insolubility in chloro-iodide of zinc and their difficult solubility in concentrated sulphuric acid. A macrochemical analysis has not yet been carried out, for obvious reasons.

14. The Mucus-globules of the Cyanophyceæ.

416. Several authors have observed globular structures in the cells of the *Cyanophyceæ*, which always lie in the peripheral part of the protoplasm (Fig 57, *s*) and are often especially abundant on both sides of the transverse walls. These may be here termed, with Schmitz, mucus-globules, although no conclusion as to their chemical composition is at present possible.

417. According to the recent investigations of Zacharias

(III, p. 12 of separata), these mucus-globules give the following reactions: They are insoluble in alcohol and in ether, and are unchanged either by boiling in distilled water or by the addition of a 1% solution of soda (Na_2CO_3). In .3% *hydrochloric acid* they swell up and become invisible, and the same result is produced by a mixture of two volumes of concentrated sulphuric acid and three volumes of water; while they swell even in a mixture of one volume of sulphuric acid and 100 volumes of water, but remain visible. 3–5% caustic potash solution causes swelling of the mucus-globules, but in a 1% solution they are unchanged. In a solution of potassium ferrocyanide acidulated with acetic acid they stand out sharply and take a vacuolar structure. In Millon's reagent they remain colorless, and also in iodine-glycerine or in chloroïodide of zinc. *Iodine and potassium iodide* solution and the above-mentioned dilute sulphuric acid cause, on the other hand, a deep brown coloring of the mucus-globules. Acetic carmine colors them deep red, and a deep stain is produced by *alum-carmine* and by Delafield's hæmatoxylin, after treatment with alcohol.

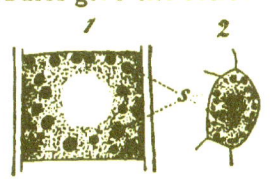

Fig. 57.—1, living cell of *Scytonema*; 2, cell of *Nostoc*, after staining with acetic carmine. *s*, mucus-globules. After Zacharias.

Their capacity for staining with hæmatoxylin has been disputed by Bütschli (I, 17). According to his statements, *eosin* is especially useful for staining them. The mucus-globules may be made visible, even in hæmatoxylin preparations, by subsequent staining with this agent. [Hieronymus (II) has studied bodies found in the cells of *Cyanophyceæ*, which he terms *cyanophycin-grains* and regards as identical with the mucus-globules of Schmitz. According to his statements, the bodies studied by him agree in their reactions with those above described, and he gives the following additional facts concerning them: They give no proteid reaction. They are quickly dissolved by nitric acid, by a solution of salt, by *eau de Javelle, chloral hydrate*, and *caustic potash* solution; but are insoluble in an artificial gastric juice, in carbon bisulphide, in acetic acid, and in a cold solution of di-sodium phosphate (Na_2HPO_4); but in a boiling concentrated solution of the last they slowly dissolve. This author states that these bodies are formed of a knot of thread, and he regards them as probably related to nuclein (cf. § 236).]

15. Tannin-vesicles.

418. In the cells of many *Zygnemaceæ* are found strongly refractive globular structures, which, as Pringsheim (II, 354) recognized, contain large quantities of tannin and are commonly termed tannin-vesicles. Such tannin-vesicles have also been observed in various Phanerogams (cf. af Klercker I, 15). But, while they are usually present in the Algæ mentioned in large numbers, and are of small size (cf. Fig. 58, *g*), in the Phanerogams they are usually single or only a few in a cell and are of a relatively large size, as in the bases of the petioles of *Desmanthus plenus* (cf. Fig. 59, *g*).

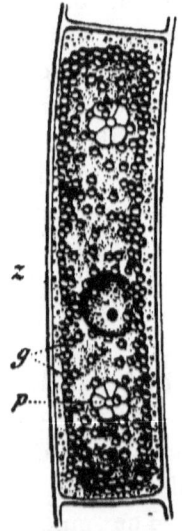

FIG. 58.—Cell of *Mesocarpus* sp. *g*, tannin-vesicles; *p*, pyrenoid; *z*, nucleus.

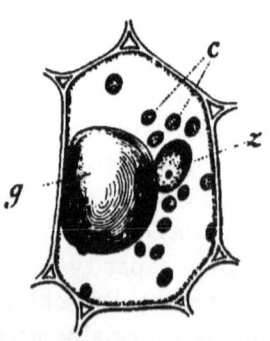

FIG. 59.—Cell from the base of the petiole of *Desmanthus plenus*. *g*, tannin-vesicle; *z*, nucleus; *c*, chloroplasts.

419. These tannin-vesicles always arise, as Klercker (I, 22) has shown, in the protoplasm, from which they are most probably separated by a true precipitation membrane of albumen tannate. Whether they contain other substances than tannins cannot at present be certainly stated; but, at all events, they cannot contain dissolved proteids, as Klercker (I, 36) has shown.

420. In the investigation of the tannin-vesicles, all the

tannin reactions described in §§ 199-208 may, of course, be used. Especially useful is Pfeffer's staining *intra vitam* with methylene blue (§ 208), which may be used with particular advantage in developmental investigations. As suitable objects for study may be suggested the cells of *Zygnema* and *Mesocarpus* (cf. Fig. 58), as well as those from the root-cap of *Pistia Stratiotes* or from the base of the petiole of *Desmanthus plenus* (cf. Fig. 59).

421. A good fixation of this coloring may be obtained by placing the stained objects in a concentrated aqueous solution of picric acid for 2-24 hours. They are then repeatedly rinsed in pure water, then placed in 10-20% alcohol, which is gradually replaced by absolute alcohol by means of Schulze's dehydrating vessel (§ 16), then transferred to a mixture of three parts xylol and one part alcohol, finally to xylol, and then mounted in balsam.* In this way I have obtained beautiful permanent preparations of *Zygnema*, in which, after six months, the tannin-vesicles alone are stained deep blue; while threads preserved for the same time in glycerine-gelatine have almost entirely lost their color.

I have not experimented as to whether a solution of ammonium picrate, recommended by Dogiel (I) for fixing methylene blue stains in animal objects, is to be preferred to the watery solution of picric acid for vegetable preparations also.

To obtain permanent preparations of the tannin-vacuoles, af Klercker (III) fixes the objects either with Flemming's chrom-osmic acid, or with a mixture of one volume of Kleinenberg's picro-sulphuric acid and one volume of a 5% solution of potassium bichromate, or with a mixture of equal parts picro-sulphuric acid and cupric acetate solution. After washing, the objects are imbedded in paraffine in the usual way. Finally, a much darker staining of the tannin

---

* Clove-oil, phenol, or aniline must not be used in the transfer to balsam, as they at once wash out the methylene blue stain.

precipitates may be obtained, by means of a feebly alkaline silver solution, in thin microtome sections.

16. The Reactions of the Various Cell-constituents.

422. In many investigations it is of interest to determine what reaction the various constituents of the cell, especially the cytoplasm and the cell-sap, have in the living cell. The reactions of the sap pressed from large pieces of tissue give very uncertain results in this respect, in view of the extensive division of labor within the cell-organism; and I therefore refrain from discussing further observations made in this way.

423. So far as the *cell-sap* is concerned, its reaction may be directly determined in those cases in which it contains a coloring matter in solution which changes its color with the reaction. Such a coloring matter is the so-called *anthocyanin* (§ 184), which appears red when the reaction is acid, blue when feebly alkaline, and green to yellow when strongly alkaline. The cell-sap is generally alkaline or neutral in blue parts of flowers, but must be acid in red parts. In the same cell, in the course of its development, a change in reaction may occur, as in the flowers of *Pulmonaria officinalis*, which are first red, and then blue. In the last-named plant, as Pfeffer has shown (V, 140), a blue color of the red parts may be produced at any time by traces of ammonia.

424. In cells with colorless cell-sap, one may reach conclusions as to its reaction by the method, proposed by Pfeffer, of introducing into it an artificial coloring matter which gives different colors, according to the reaction. According to Pfeffer (II, 266), *methyl orange* is especially suited to this purpose, as its orange-yellow color is not changed by dilute alkalies. Pfeffer used in his experiments a .01% solution. Pfeffer has (II, 259 and 267) also experimented with cyanin, tropæolin and corallin; but these dyes proved less useful.

425. It is, in most cases, more difficult to determine the reactions of the *protoplasm*, which never contains coloring matters that show the reaction directly.

In the large plasmodia of *Æthalium septicum* Reinke (I, 8) was able to determine the alkaline reaction macroscopically; and he deduced the presence of a volatile alkali from the observation that a bluing of litmus-paper occurred when it was not in direct contact with the plasmodium. But this observation does not in any way exclude the presence of other alkaline substances in the plasmodium; and it has been shown to be probable, by Schwarz (I, 33), that the alkaline reaction of the protoplasm in the higher plants probably does not depend on the presence of ammonia or ammonium compounds.

426. In these plants Schwarz (I, 20) attempted to determine the reaction of the protoplasm by killing cells that naturally have a colored cell-sap by alcohol, heat, crushing, or electricity, and then noting the color assumed by the dead protoplasm. Again, he treated colorless cells in the same way in a feebly acidified extract of *borecole* leaves which is yellow-red, purple-red, or red-violet in an acid solution, violet when neutral, and blue, blue-green, grass-green, yellow, or yellowish orange when feebly alkaline. Schwarz found that, after killing by electricity, the protoplasm of a few cells was blue-green; but in most, it was blue-violet, or red-violet, and he therefore concluded that the reaction of the protoplasm is alkaline.

On the other hand, Arthur Meyer (III) observed that the coloring matter of kale is not violet, but blue, when the reaction is neutral, and that, when tin-foil electrodes are used in conducting the current, a violet tin compound of the coloring matter is formed and taken up by the dead protoplasm, and that various colorations of the extract may be caused near the electrodes by the decomposition of the salts contained in it. There can therefore be no doubt that Schwarz's method cannot give trustworthy results.

427. The most certain conclusions may be reached by Pfeffer's methods of introducing artificially certain colored

indicators into the living cell. In fact, Pfeffer has (II, 259 and 266) already shown the alkaline reaction of the cytoplasm in various cells by the aid of *cyanin* and methyl orange.

17. Plasmolysis (Plasma-membranes).

428. If living plant-cells are placed in a solution of salt or the like, the protoplasm withdraws from the wall in the form of a continuous sac, if the solution be above a certain degree of concentration, in consequence of its power of taking up water. This proceeding, now generally known as plasmolysis, offers the best means of showing the continuity of the protoplasmic body in cells poor in protoplasm, and it also plays an important part in various morphological and physiological researches.

429. Concerning the media used in plasmolysis, it is especially important that they shall exert no unfavorable influence on the cells in the degree of concentration in which they are employed. They should also have neither a markedly acid nor an alkaline reaction. It is best, then, to use neutral salts like saltpeter (KNO_3) or salt (NaCl), or organic compounds like cane-sugar or glycerine. The latter has the advantage, in some cases, of exerting a clearing effect in consequence of its higher refractive index. No general directions can be given as to the concentration of the solutions to be used, and in the choice of the proper concentration the *isotonic coefficient* * of the compound used should be especially regarded. But, in general, clearly visible plasmolysis may be obtained by using a 4% solution of saltpeter or a 15% solution of cane-sugar.

430. If the objects to be plasmolyzed are not prescribed by the nature of the investigation, it is best to choose such objects as are most adapted to the observation of plasmolysis, and as have the greatest power of resisting the injurious influences connected with their preparation. Thus the cells of *Spirogyra* furnish very suitable material for the demonstration of plasmolysis. Such cells as naturally contain a

---

* [Cf. Zimmermann I, 199-201 : also below, p. 263.]

colored cell-sap are also very useful, as, for example, the epidermal cells of the red lower surface of the leaf of *Tradescantia discolor*. The earliest beginnings of plasmolysis are easily observed in such cells.

431. In many cases the plasmolysis may be made plainer by an artificial coloring of the cell-sap. In cells containing tannic acid this may be accomplished by exposing the objects on the slide, in a drop of the plasmolyzing solution, to the vapor of *osmic acid* for a time, according to the method described in § 308. The protoplasts are then completely fixed in their original position and are much easier to observe on account of the browning or blackening of the cell-sap. Plasmolysis may also often be made plainer by preliminary staining *intra vitam*.

Finally, by adding an indifferent pigment, like eosin, to the plasmolyzing fluid, a difference in color between it and the cell-sap may be produced.

To test for the presence of a living plasma-body within the *vessels*, Th. Lange (I, 404) injected large pieces of tissue with a 5% solution of saltpeter under the air-pump, and then added a dilute solution of picric acid to fix the tissues. After washing in water and dehydrating in alcohol, he placed them in clove-oil. Very good thin sections were prepared from the material so treated, and in these the protoplasm was stained with borax-carmine or eosin.

Abnormal Plasmolysis.

432. As was shown by H. de Vries (I), many solutions cause the complete killing of the protoplasm and its inclusions, with the exception of the inner plasma-membrane bordering on the cell-sap, which is not changed in its osmotic relations, so that it still completely excludes the cell-sap (cf. Fig. 60). If this proceeding, which de Vries calls abnormal plasmolysis, does not justify such far-reaching conclusions as he has drawn from it (cf. on this point Pfeffer VIII, 240), yet abnormal plasmolysis may do good service in many investigations.

433. To produce it, one may use a 10% solution of saltpeter to which a little eosin has been added. This pigment has the double advantage of at once staining the killed part of the protoplasm and of causing a sharper definition of the uncolored cell-sap. Pretty large *Spirogyra*-cells furnish excellent objects for study (cf. Fig. 60).

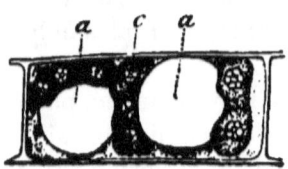

FIG. 60.—Cell of *Spirogyra* in ab-normal plasmolysis with a 10% saltpeter solution containing eosin. *aa*, the two isolated vacuoles; *c*, remains of the dead chloroplasts.

434. For fixing the isolated vacuolar membranes Went recommends (I, 314) $\frac{1}{2}$–1% chromic acid, which he allows to act one or two days. He then washes the objects in running water six hours and transfers them, in the usual way, to paraffine, for the preparation of microtome sections.

18. Methods of determining whether certain Bodies lie in the Cytoplasm or in the Cell-sap.

435. This question often cannot be answered by direct microscopical examination, and already various methods have been devised which make a certain decision possible even in difficult cases. Wakker (I) observed, with microscope tipped down, what motions the bodies in question underwent with the slide in a vertical position. If they simply fall downward in the cells in a vertical line in consequence of their greater specific gravity, it is very probable that the bodies lie in the cell-sap. But, since the starch-bearing chromatophores of the starch-sheath, which undoubtedly lie in the cytoplasm, sink down rather rapidly in the cells, as was recognized by Dehnecke (I, 9) and Heine (I, 190), it has seemed useful to me to modify this method by reducing the ready displaceableness of the protoplasm by killing the cells, which may be easily done with an iodine and potassium-iodide solution. The movement of the chromatophores in the starch-sheath was at once stopped by iodine solution, while the protein crystalloids in the epidermis of the leaf of *Polypodium irreoides*, which undoubtedly lie in the

cell-sap, continued the motions due to their weight after the same treatment (cf. Zimmermann II, 68).

436. Wakker (I) has also used abnormal plasmolysis with a 10% solution of saltpeter, containing eosin (cf. §§ 432 and 433), for the same purpose. It is often to be determined with certainty by direct observation whether the bodies in question lie in the vacuoles isolated by this method. Besides, a further confirmation of the conclusions reached by direct observation may be obtained from the movements taking place in such preparations when the slide is placed vertically.

19. Aggregation.

437. Complicated changes of arrangement take place within the cells of the glandular hairs of *Drosera rotundifolia* in consequence of chemical stimulation, which consist essentially in the origin of rapid circulation-currents in the protoplasm and the breaking up of the large central vacuole into a large number of small vacuoles which gradually contract more and more. This process, discovered by Darwin and studied in detail especially by H. de Vries (II), will be termed exclusively, in the following account, aggregation, in agreement with de Vries, though Darwin and various other authors use the same term also for the artificial precipitation which accompanies the process, but also occurs in very many plants.

438. The tentacles of *Drosera* are especially adapted to the observation of aggregation, since their vacuoles stand out very sharply on account of their red cell-sap. Aggregation may be produced in them by bringing a leaf of the plant in contact with an insect, a bit of cooked albumen, or the like. It may also be observed in isolated tentacles which have been placed in a 1% solution of ammonium carbonate. Aggregation then begins at the bases and at the tips of the tentacles, and may be especially well followed in its separate stages at their middles (cf. Fig. 61).

439. According to recent investigations of Bokorny (IV), who observed aggregation in other plants also, it is produced by the most various *basic* substances; but if they have highly poisonous properties, as caustic potash or ammonia, they must be used in very dilute form. This author observed a very far-reaching aggregation on leaving superficial sections from perianth-leaves of *Tulipa suaveolens* for two or three days in a .1% *coffein* solution.

FIG. 61.—Cell from a marginal tentacle of the leaf of *Drosera rotundifolia* in three different stages of aggregation, caused by a .1% solution of ammonium carbonate. The interval between I and II is six minutes, and that between II and III, two minutes. After H. de Vries.

440. A somewhat different appearance, which is certainly very closely related to aggregation, was observed by Bokorny (IV, 451) on placing sections of the margin of the stigma of *Crocus vernus* in a .1% solution of coffein. The vacuolar wall did not draw together as in normal aggregation, but the whole protoplasmic mass contracted, so that there arose between it and the cell-membrane a space filled with water.

20. Artificial Precipitates.

441. Artificial precipitates may be produced in the cytoplasm and in the cell-sap without injury to the vitality of the cell, by the most various reagents. Such excretions may often arise in the cell-sap during plasmolysis. The chemical composition of these bodies has as yet been determined in but few cases; but it may be regarded as very probable that *protein*-like compounds, especially, are widely distributed in these precipitates with *tannins* and related substances.

442. Precipitates which consist of relatively pure *tannic acid* are produced in a great variety of cells by alkaline carbonates (cf. § 206). That these usually globular bodies contain no important amount of proteïd substances was recognized by Klercker (I).

Similar precipitates were produced by Bokorny (V and IV) in various plant-cells by the most various basic compounds, as, for example, by a .1% solution of *coffein*. These may lie partly in the cytoplasm and partly in the cell-sap, and, since they also occur in cells free from tannin, cannot consist of tannin in all cases, at least, But to what extent proteid substances or " active albumen " (cf. § 445), which forms their chief constituent, according to Bokorny, occur in them cannot be certainly determined from our present knowledge. It is not impossible that we have here to do with very various bodies.

443. Precipitates occurring in *plasmolysis* were first observed by Pfeffer (II, 245) in the root-hairs of *Azolla*, and they occurred in the same manner whether the plasmolysis was accomplished by means of sugar, saltpeter, or calcium chloride. These precipitates agree essentially with those produced by ammonium carbonate, and consist chiefly of tannin, in most cases. But, as af Klercker (I, 29) has shown, other substances must be present in the cell-sap which remain dissolved in the vacuolar fluid during the excretion, and prevent the partial re-solution of the tannin. It is also noteworthy that Klercker (I, 43) has observed similar precipitates in *artificial cells* of tannate of glue, on plasmo- lyzing them with a solution of saltpeter, while plasmolysis with sugar contracted the whole cell into a glassy mass.

444. Plasmolytic precipitates which are not due to the presence of tannins may be very easily ob- served in the epidermal cells of the red under surface of the leaves of *Tradescantia discolor*. If tangential sections of these cells be placed in a small dish with a 10% solution of saltpeter, it may be seen in ten minutes that strong plasmo- lysis has taken place in all the cells, and pretty strongly refractive, deeply colored spheres (cf. Fig. 62) occur in most cells, while the cell-sap has in some cases become markedly clearer. If water is afterwards added, these precipitates are redissolved. No precipitates are produced in these cells by ammonium carbonate, and no blackening occurs with osmic acid.

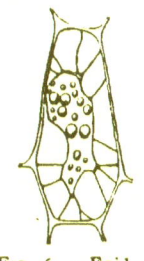

Fig. 62. — Epider- mal cells from the under side of the leaf of *Tra- descantia discolor* plasmolyzed with a 10% saltpeter solution, show- ing plasmolytic precipitates.

21. The Loew-Bokorny Reagent for "Active Albumen."

445. Loew and Bokorny (cf. I and II) have, in various papers, maintained the view that living and dead protoplasm differ from each other in that only the former has the power to precipitate silver from an *alkaline silver-solution*. These authors conclude from this that living albumen contains aldehyde groups which at once undergo a breaking up, at its death. [They have observed a precipitation, in living cells treated with a .1% solution of coffein, of abundant small granules which they believe to consist of "active albumen," and term "proteosomes."]

446. It has, however, been shown by various authors (cf. especially Pfeffer VII) that the observations of Loew and Bokorny are partly incorrect and that the bodies called by them "active albumen" certainly consist in part of tannin and similar substances. But, since the Loew-Bokorny "*Life-reagent*" has already been used by other investigators for the recognition of living albumen, and may perhaps be capable of furnishing a basis for some conclusions concerning the contents of the cell, when more critically employed, the methods used by these investigators may be briefly outlined here.

447. The silver reagent called by Loew and Bokorny "*Solution A*" is prepared by mixing 13 ccm. of caustic potash solution of specific gravity 1.33 (containing $33\frac{1}{2}$% of KOH) and 10 ccm. of aqua ammonia of specific gravity .96 (containing 9% of NH_3) and diluting the mixture to 100 ccm. For use, 1 ccm. of this solution is mixed with 1 ccm. of a 1% silver nitrate solution, and the mixture is diluted to one liter (1000 ccm.).

448. The "*Solution B*" is prepared by adding 5 to 10 ccm. of a saturated solution of lime to a liter of. a $\frac{1}{1000}$% solution of silver nitrate.

Both solutions must be used in large quantities on account of their extreme dilution, and but a small number of the objects used should be placed in them. The deposition of silver usually begins only after some hours, and it is generally necessary to leave the objects from 5 to 24 hours in the solutions.

A more rapid reaction is obtained with a more concentrated solution containing 1 gram AgNO, and .3 gram NH, to a liter of water. This can be used, however, only with resistant objects and with such as contain neither sugar nor tannin (cf. Bokorny VI, 195).

22. Protoplasmic Connections.

449. Proof has been furnished by the investigations of Tangl, Gardiner, Russow, and others that, besides the elements of the sieve-tubes, the protoplasts of the separate cells of various other tissue-systems are in direct connection with each other through perforations of their cell-walls. Kienitz-Gerloff (I, 22) has recently concluded from his researches that all the living elements of the entire body of the higher plants are united by protoplasmic threads. The threads which accomplish this union, the so-called protoplasmic connections, are, however, in most cases so fine that nothing can be seen of them within the living cell; and rather complicated preparation-methods are necessary to their recognition.

450. If one has to do with relatively *thick* protoplasmic connections, like those of many sieve-tubes, they may be made visible, in many cases, simply by treatment with chloroïodide of zinc or with iodine and sulphuric acid. Gardiner (II, 55, note 4) recommended for this purpose a solution of Hofmann's violet in concentrated sulphuric acid. This solution has a brownish color and does not stain strongly sections placed in it. But if the acid is washed out with a large quantity of water, after acting for about half a minute, the sections take at first a green, then a bluish, and finally a violet color, and the protoplasts are almost exclusively colored. I have obtained in this way very instructive preparations of alcoholic material, adding to pretty thin sections on the slide, after drying them externally with filter-paper, a drop of the sulphuric acid containing the staining material, and at once dropping on a cover-glass to prevent too great warping of the sections. After a short time the whole slide is plunged into a large vessel of water in which it remains until the sections have become clear violet. Although the cover-glass usually separates from the slide, some sections generally remain attached to it, and these can, naturally, best be used for study.

451. A very deep staining of the protoplasmic connections contained in the sieve-pores may be obtained in microtome sections of stems of *Cucurbitaceæ* by the use of Altmann's acid fuchsin method (cf. § 345). This brings out the protoplasmic threads without the previous use of any swelling media.

In many cases the double staining, described in § 290, with *aniline blue* and *eosin* gives very fine preparations of the sieve-plates with the protoplasmic connections passing through them.

452. *Delicate* protoplasmic connections, on the other hand, have as yet been made visible only by the successive action of swelling and staining media.

For swelling, chloroïodide of zinc and sulphuric acid have been chiefly used.

Chloroïodide of zinc was recommended especially by Gardiner (II). He treated sections of fresh material first with a solution of iodine and potassium iodide, then added chloroïodide of zinc and let it act a longer or shorter time, according to the capacity of the membranes for swelling, commonly about 12 hours. Before staining, the chloroïodide was washed out with water or, where the membranes were much swollen, with alcohol.

453. When sulphuric acid is used, the sections are also usually first fixed with iodine and potassium iodide solution. Kienitz-Gerloff (I, 8) used for this purpose a solution containing .05 gram of iodine and .20 gram of potassium iodide in 15 grams of water. He recommends, especially for juicy tissues, the method used by A. Fischer (III) with the best results in the study of the contents of the sieve-tubes (cf. § 455). This author scalded large plants or large parts of plants in boiling water as quickly as possible and hardened them, before cutting, in absolute alcohol.

Concerning the strength of the sulphuric acid used for swelling it may be remarked that it has been used partly in concentrated form, partly after dilution with one fourth its volume of water. With membranes which swell strongly, the dilute acid is, of course, to be preferred. The time for it to act depends chiefly on the character of the membranes. But usually a few seconds is sufficient for concentrated acid. The acid must, of course, be removed by careful washing in water, before staining.

453. The following dyes have been used for staining sections treated in this way.

1. *Hofmann's blue* (same as aniline blue). Gardiner (II) recommends a solution of this dye in 50% alcohol saturated with picric acid. The solution is washed out with water, and the preparation is then either mounted in glycerine, or is gradually transferred through dilute to strong alcohol, cleared with clove-oil, and mounted in Canada balsam. In this way very permanent preparations of the protoplasmic connections may be obtained.

Gardiner also used a solution of Hofmann's blue in 50% alcohol acidified with a few drops of acetic acid. The subsequent treatment is the same as in the previous case.

Terletzki (I, 455) stains simply with a strong aqueous solution of aniline blue and examines the preparation in water, after washing.

2. *Hofmann's violet* is used by Gardiner, simply in an aqueous solution. This at first stains wall and protoplasm about equally; but, after lying for a long time in glycerine, in many cases several days, the color is removed from the walls, while the protoplasts and protoplasmic connections remain strongly stained.

3. *Methyl violet* is recommended by Kienitz-Gerloff (I, 9) for such cells, like those of hairs, etc., as do not permit the penetration of Hofmann's blue, on account of cuticularization of the walls. He uses it in a concentrated aqueous solution.

It may be observed that Gardiner finds it useful to brush over the sections with a camel's-hair brush after swelling and staining.

454. A method differing essentially from previous ones has lately been used by Kohl (I, 12) for demonstrating the protoplasmic connections. It agrees in essence with Loeffler's method for staining cilia (§ 476). After preliminary mordanting with tannin, Kohl stained with *methylene blue* or Bismarck brown; and if the staining of cell-walls or gelatinous sheaths, due to the presence of pectic substances, interfered with observation, these substances were removed by an acid. This author has not given more exact details of his method.

23. Contents of Sieve-tubes.

455. Since the contents of the sieve-tubes, which communicate by relatively large openings, are partly pressed out on cutting the tissues, and undergo the most various changes, A. Fischer (III) has devised a method of fixing the

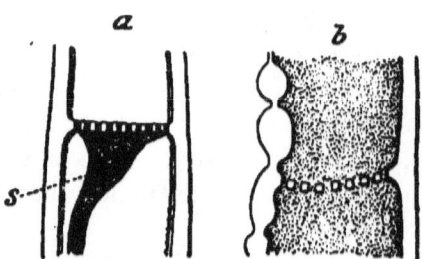

FIG. 63.—Parts of sieve-tubes of *Cucurbita Pepo*. *a*, from a plant cut and then scalded; *s*, "Schlauchkopf". *b*, scalded in an uninjured condition (× 675). After Fischer.

contents of the sieve-tubes in the uninjured tissues. For this purpose he plunges carefully unpotted plants or pieces of an uninjured plant, such as branch-tips, into *boiling* water and leaves them in it until the contents of the sieve-tubes are coagulated, for which two to five minutes' exposure is usually quite sufficient. He was able to show that the accumulations of strongly refractive and readily staining substance observed at one side of the sieve-plate in *Cucur-*

bita and many other plants when the ordinary methods are employed, the so-called "Schlauchköpfe" (Fig. 63, *a, s*), represent artificial products and originate only when the sieve-tube system is cut. They were entirely wanting in material scalded as above (Fig. 63, *b*), but were not changed by boiling water in tissues where they had arisen from cutting before the scalding.

456. According to investigations of this scalded material by A. Fischer (VI), we may distinguish three sorts of sieve-tubes, according to the organization of their contents, as follows:

1. Sieve-tubes with *coagulable contents* (in the *Cucurbitaceæ*) have a thin protoplasmic wall-layer and a clear sap coagulable by heat.

2. Sieve-tubes with *mucus* (as in *Humulus*) contain a delicate wall-layer filled with large and small slime-masses and a clear, non-coagulable, watery fluid.

3. Sieve-tubes with *starch-grains* (as in *Coleus*) contain a delicate wall-layer carrying small masses of mucus and a clear, non-coagulable fluid with small starch-grains.

Appendix.

METHODS OF INVESTIGATION FOR BACTERIA.

457. Since the methods employed in Bacteriology differ in many respects from those used in the study of other plants, I have preferred to collect the former into a special chapter. In the following pages I have not attempted to give an even approximately exhaustive compilation of bacteriological methods. I have rather chosen to bring together a number of trustworthy methods of preparation which should be quite sufficient for most cases in the study of Bacteria. I must refer persons who wish to devote themselves especially to Bacteria to the special works on the subject, particularly those of Günther (I) and Hueppe (I).

I. The Observation of Living Bacteria.

458. The observation of living Bacteria may be generally conducted like that of other lower organisms. If it is to be continued for a long time, the Bacteria may best be placed in a hanging drop (cf. § 2). Under some circumstances, a frequent renewal of the culture-fluid is necessary.

In case of rapidly motile Bacteria, they may be brought to rest by proper fixing media, like the fumes of iodine or of osmic acid.

Finally, I may remark that, in some cases, the dark-field illumination of the Abbé condenser may be used with success in the examination of Bacteria.

II. Fixing Methods.

1. Cover-glass Preparations.

a. Fixing by Dry Heat.

459. For fixing Bacteria from fluids containing them, from gelatine cultures, or the like, dry heat is almost exclusively used at present, and commonly, according to R. Koch's method, as follows :

A small drop of fluid containing Bacteria is transferred, by means of a platinum wire sterilized by heat, to a carefully cleaned cover-glass,* and spread out over it as evenly as possible. If the Bacteria are taken from a solid substratum, a drop of water is first placed upon the cover and a very small particle of the material is rubbed up in it as completely as possible with a platinum wire.

The fluid is now allowed to evaporate from the cover-glasses thus smeared with Bacteria, at the ordinary temperature, until the preparation is "air-dry."

The Bacteria are then fixed by passing the cover-glasses, with their Bacteria-sides upward, three times through the non-luminous flame of a Bunsen burner.† Johne's rule may give an idea of the rapidity with which this should be done. According to this, the hand should describe an horizontal circle a foot in diameter in a second, moving at an equal rate throughout the course, and passing through the flame at one part of it.

460. In this way the Bacteria are so fastened to the cover-glass that they may be treated with staining-fluids and other media without fear of separation. It is thus possible to stain or restain, at any time, preparations which have been mounted for a long time in Canada balsam. For this pur-

---

* The cleaning of cover-glasses may be accomplished by the method proposed by Günther (I, 40, note), which consists in heating the glasses, after cleaning with alcohol, in the non-luminous flame of a Bunsen burner.

† In these and the following manipulations the so-called Kuehne forceps are very convenient. Their arms are bent at about 1½ cm. from their tips, and end in broad surfaces.

pose one need only gently warm the preparation until the balsam becomes fluid, remove the balsam from the raised cover-glass with xylol, and finally wash off the xylol with alcohol.

It may be remarked, finally, that Bacteria fixed in the above manner may be preserved in this condition for an indefinite time without harm, if they are protected from dust and moisture by being wrapped, for example, in filter-paper.

461. To remove the strongly staining substance from the red blood-corpuscles, in preparations from the blood, Günther (I, 63) recommends rinsing the objects, fixed in the usual way, in 1–5% *acetic acid*. The stainable hæmoglobin is thus extracted from the red corpuscles, and a large part of the blood-plasma is washed out of the preparations, leaving the Bacteria unchanged. Preparations which give no satisfactory results with this method, on account of having been kept dry for a long time, have been treated by Günther with a 2–3% solution of pepsin, with the best results.

b. Other Fixing Methods.

462. Since certain inequalities can hardly be avoided with the fixing methods described above, H. Moeller (II, 274) has proposed fixing the air-dry preparations with absolute *alcohol*, instead of heating them; he leaves them in this fluid two minutes.

463. A. Fischer (II) demonstrated the noteworthy fact that artificial appearances often arise, especially when preparations are allowed to dry, which are chiefly the result of plasmolysis of the bacterial cells (cf. § 428). As Fischer has shown, the Bacteria are plasmolyzed by solutions of pretty slight concentration. In general a 1% salt solution is sufficient to produce plasmolysis in most Bacteria. Fischer (II, 73) recommends the use of a 10% solution of *lactic acid* for fixing Bacteria, which does not prevent subsequent staining with alcoholic solutions of aniline dyes.

464. Besides, the fixing methods used for higher plants

may certainly be used with some success in the investigation of the minute structure of the bacterial cell. Their use may generally be successfully carried out by Overton's method. But it should be noted that the membranes of the Bacteria are characterized by relatively great impermeability. It has been shown by A. Fischer (II, 72) that 1% osmic acid and a 1% solution of corrosive sublimate, in particular, cause only an incomplete fixation of bacterial cells.

2. Sections.

465. Absolute alcohol is commonly used for fixing Bacteria within infected organisms, pieces of the tissue being placed directly in it. The microtome should be used for cutting sections from these, after imbedding in paraffine (§ 43) or celloidin (§ 49a).

III. Staining Methods.

466. Under this head a number of methods will be described which can be successfully used, in most cases, for recognizing the presence of Bacteria in a fluid or in a diseased organism. Then follows a special description of staining methods for tubercle Bacilli, the spores, and the cilia of Bacteria.

1. Staining with Loeffler's Methylene Blue.

467. Loeffler's methylene blue consists of 30 ccm. of a concentrated alcoholic solution of methylene blue and 100 ccm. of an aqueous .01% solution of caustic potash. It keeps indefinitely.

With *cover-glass* preparations, it may be used by carefully warming the preparation with a few drops of the stain until steam is seen to rise, then washing off the stain with water and drying in the air, without heat. A drop of Canada balsam is then placed on a slide, and the cover-glass is placed on it. This method does not give a very deep stain, but often brings out delicate differentiations sharply.

For staining *sections* which would be injured by warming, the methylene blue may be allowed to act for a longer time. The staining fluid is then washed off with water, and the preparation is transferred to balsam, either by being first allowed to dry (§ 23), or by the use of aniline between water and xylol (§ 24). Many Bacteria, such as the anthrax *Bacillus*, endure treatment with alcohol very well, and may therefore be transferred to balsam in the ordinary way.

2. Ziel's Carbol-fuchsin.

468. Ziel's solution of carbol-fuchsin is prepared by rubbing up one gram of fuchsin with 100 ccm. of a 5% aqueous solution of carbolic acid, with the gradual addition of 10 ccm. of alcohol. It is very stable.

With *cover-glass* preparations, it is allowed to act only about a minute. Under some circumstances its action may be hastened by warming. The preparations may be washed in water and mounted, after drying, in Canada balsam. But strongly stained objects will endure longer washing with alcohol, and may be transferred to xylol, and then to Canada balsam.

Carbol-fuchsin seems less adapted to use with sections.

3. Ehrlich's Aniline-water Solutions.

469. These solutions are prepared by adding 11 ccm. of a concentrated alcoholic solution of fuchsin, gentian violet, or methyl violet to 100 ccm. of aniline-water. Turbidity arises at first in this mixture, which prevents its immediate use. But it may be used for staining in 24 hours, after previous filtering. It remains fit for use but a few weeks.

It is sufficient to let these solutions act for a minute, while heated, on *cover-glass* preparations. The dye is then washed off with water, and the preparation is dried and mounted in balsam.

Better staining of the Bacteria contained in sections is obtained by Gram's method, described in the next paragraph.

4. Gram's Method.

470. The so-called Gram's method is adapted especially for *sections*, because it stains the Bacteria in them deeply without staining the nuclei at the same time. But it may also be used with cover-glass preparations, especially if they contain many other stainable bodies besides Bacteria.

According to Gram's original account, this method consisted in placing the sections first for several minutes in Ehrlich's aniline-water-gentian-violet solution (§ 469), and then transferring them to a solution containing one part of iodine and two parts of potassium iodide in 300 parts of water. After a few minutes in this, they are washed with alcohol until no more color comes off, then transferred to clove-oil, which removes more of the dye, and finally mounted in balsam.

471. But, according to Günther (I, 89), it is better, in most cases, to treat the sections, after removal from the iodine solution, for half a minute with alcohol, then *just* ten seconds* with 3% hydrochloric acid-alcohol, and then at once with pure alcohol until they are completely decolorized. For transferring them from alcohol to balsam, this author recommends xylol, instead of clove-oil.

According to the method described by Weigert, aniline is gradually dropped upon the sections, differentiating and dehydrating them, and they are then passed through xylol into balsam.

472. A sharp double staining, by which the nuclei are differently stained from the Bacteria, may be obtained by preliminary staining with picro-carmine (§ 318). This solution is allowed to act one or two minutes on the sections, which are then carefully washed with water, placed in alcohol, and finally stained again according to the Gram or the Gram-Günther method.

---

* For pretty thin paraffine sections this time is certainly too long.

5. Staining Tubercle-Bacilli.

473. The *Bacilli* of tuberculosis and of leprosy are characterized by a peculiar behavior with staining media which makes possible the staining of them alone in a mixture of Bacteria, and therefore their certain distinction from other species of Bacteria. Of the numerous methods recommended for staining tubercle *Bacilli*, only the following, due to Czaplewski (I), need be here referred to; and I have obtained excellent results with it.

Cover-glass preparations are treated, after fixing, first for a minute with carbol-fuchsin (§ 468) heated to boiling. They are then washed with the so-called Ebner's fluid * until hardly a trace of color can be seen, are then repeatedly rinsed with pure alcohol, and stained again with a mixture of three parts water and one part concentrated alcoholic solution of methylene blue. This is then rinsed off with water, and the preparation is dried and mounted, in the usual way, in balsam.

If a mixture of tubercle *Bacilli* and other Bacteria, for example, which can easily be prepared by mixing pure cultures, be treated in this way, it will be found that, with the exception of the very rare lepra-*Bacilli*, only the tubercle-*Bacilli* are colored red, while all other Bacteria are blue.

474. The staining of tubercle *Bacilli* in *sections* can be accomplished by essentially the same methods. If one has paraffine sections attached to the slide with albumen (§ 52), they may be heated in carbol-fuchsin, whose action for a few minutes is sufficient. But if the sections will not endure heating, the fuchsin solution must be allowed to act for a longer time (about 24 hours). It is also better to wash the methylene blue from sections with alcohol, and to transfer them to balsam through xylol.

In preparations treated in this way, only any tubercle *Bacilli* that may be present are stained red; all other Bacteria, and the nuclei of the tissue, are blue.

---

* This consists of 20 parts water, 100 parts alcohol, .5 part hydrochloric acid, and .5 part sodium chloride.

6. Staining the Spores of Bacteria.

475. A well-differentiated staining of the spores of Bacteria may be obtained, according to the method proposed by H. Moeller (II), by plunging cover-glass preparations fixed by heat or by alcohol in 5% chromic acid * for from five seconds to ten minutes, then thoroughly rinsing in water, adding carbol-fuchsin in drops, and warming the whole in the flame for 60 seconds, allowing it to boil up once. The carbol-fuchsin is then poured off, the cover-glass is plunged in 5% sulphuric acid until it is decolorized, and again thoroughly washed with water. An aqueous solution of methylene blue or malachite green is then allowed to act for 30 seconds, and is rinsed off with water. The preparation is allowed to dry and mounted in balsam. In good preparations, the spores are to be seen as bright red spots within the blue or green bodies of the Bacteria.

Zinc chloride or chloroïodide may be used as a mordant, instead of chromic acid, but these commonly require a longer time of action than the latter.

7. Staining the Cilia of Bacteria.

476. For staining the cilia of Bacteria, Trenkmann (I) and Loeffler (I) have proposed different methods. According to Loeffler's latest publication, the following method is best adapted for the purpose. The Bacteria are first spread upon the carefully cleaned cover-glass (§ 459, note), and fixed by being passed three times through the flame, too strong heating being carefully avoided. The mordant is then placed on the warm cover-glass. This is best prepared

---

* In general, the action of chromic acid for about 30 seconds is sufficient; but different species of Bacteria show great differences in this respect. According to Moeller, the most favorable duration of its action is: for the brown potato-*Bacillus*, 30 sec.; for the yellow one, 2 min.; for the white one, 10 min.; for *Bacillus cyanogenus*, 30 sec.; for the anthrax *Bacillus*, 2 min.; for the tetanus *Bacillus*, 2 min. I obtained beautiful staining of the spores of *Bacillus subtilis* on allowing the chromic acid to act 30 sec. on cover preparations fixed by heat.

by mixing 10 ccm. of a 20% aqueous tannin solution, 5 ccm. of a cold saturated solution of ferrous sulphate, and 1 ccm. of an aqueous solution of fuchsin. According to the character of the Bacteria, a few drops of sulphuric acid or of caustic soda* must be added to this mixture. The cover-glass holding it is now warmed over the flame until steam is formed and, after half a minute to a minute, the mordant is rinsed off with distilled water, and the cover-glass is dried as usual. Then the staining fluid is dropped on until the cover-glass is wholly covered by it, the whole is again warmed for a minute until steam forms, and is finally rinsed in a stream of water, dried, and mounted in balsam in the usual way. Loeffler recommends as a staining-fluid, a solution of fuchsin in aniline-water, or a mixture of 100 ccm. of aniline-water, 1 ccm. of a 1% soda solution, and solid fuchsin in excess.

---

* For staining the cilia of typhus *Bacilli*, 22 drops of a 1% aqueous solution of sodium hydrate should be added to 16 ccm. of the above mordant; for *Bacillus subtilis*, 28 to 30 drops. On the other hand, the cholera *Bacilli* require the addition of ½ to 1 drop of a sulphuric acid that will just neutralize the same volume of 1% caustic soda; *Bacillus pyocyaneus*, the addition of 5 to 6 drops of the above acid; *Spirillum rubrum*, 9 drops of the same. But the above mordant has just the right reaction for *Spirillum concentricum*. According to the writer's experiments, it is also well suited to *Spirillum Undula* without further addition.

TABLES FOR REFERENCE.

Weights, Measures, and Temperature.

THE *metric system* is based on the *meter*, which was intended to be one ten-millionth of a meridian quadrant of the earth in length.

The unit of capacity is the *liter*, whose volume is that of one cubic decimeter or 1000 ccm.

The unit of weight is the *gram*, which is the weight of one cubic centimeter of water at 4° C.

The *centigrade* or Celsius' thermometer has for its zero the freezing-point of water, and for its 100° point the boiling-point of water. One degree of the scale is $\frac{1}{100}$ of this interval.

Comparison of Measures of Length.

English and U. S.		Metric.	
1 foot	=	.3048 meter	= 30.48 cm.
1 inch	=	.0254 "	= 25.4 mm.
$\frac{1}{8}$ "	=	3.175 mm.	
$\frac{1}{100}$ "	=	.254 "	= 254. μ
$\frac{1}{1000}$ "	=	.0254 "	= 25.4 μ
39.37 in.	=	1 meter	
.3937 in.	=	1 cm.	
.0394 in.	=	1 mm.	

Comparison of Measures of Capacity.

English.		U. S.	Metric.
1 quart, Imp.	=	1.2 qt., wine	= 1.135 liter.
.833 qt., "	=	1 qt., "	= .9463 "
1 fluid ounce	=		28.38 ccm.
1 " dram	=		3.55 "
.8811 quart, Imp.	= 1.0567 qt., wine		= 1 liter
.0352 fl. oz., or .2817 fl. dr.			= 1 ccm.

Comparison of Weights.

Avoirdupois.		Apothec.		Metric.
		1 grain	=	.0648 gram.
1 dram	=	.4558 dr.	=	1.7718 "
2.194 dr.	=	1 dram	=	3.888 "
1 ounce	=	.9115 oz.	=	28.3495 "
1.0971 oz.	=	1 ounce	=	31.1035 "
1 lb.	=	1.215 lb.	=	453.59 "
.8229 lb.	=	1 lb.	=	373.242 "
.5643 dr.	=	15.432 gr.	=	1 gram.

Comparison of Thermometer Scales.

$t° C. = \frac{5}{9}(t - 32)° F. \quad t° F. = \frac{9}{5}t° C. + 32°.$

°F.	°C.	°F.	°C.	°F.	°C.	°C.	°F.	°C.	°F.	°C.	°F.
400	204.5	130	54.4	45	7.2	300	572	110	230	35	95
350	176.7	120	48.9	40	4.4	280	536	100	212	30	86
300	148.9	110	43.3	35	1.7	260	500	95	203	25	77
280	137.8	100	37.8	30	— 1.1	240	464	90	194	20	68
260	126.6	95	35.0	25	— 3.9	220	428	85	185	15	59
240	115.5	90	32.2	20	— 6.7	200	392	80	176	10	50
220	104.4	85	29.4	15	— 9.4	190	374	75	167	5	41
200	93.3	80	26.7	10	— 12.2	180	356	70	158	0	32
190	87.8	75	23.9	5	— 15.0	170	338	65	149	— 5	23
180	82.2	70	21.1	0	— 17.8	160	320	60	140	— 10	14
170	76.7	65	18.3	— 5	— 20.5	150	302	55	131	— 15	5
160	71.1	60	15.5	— 10	— 23.3	140	284	50	122	— 20	— 4
150	65.5	55	12.8	— 15	— 26.1	130	266	45	113	— 25	— 13
140	60.0	50	10.0	— 20	— 28.9	120	248	40	104	— 30	— 22

Specific Gravity and Percentage Composition of Solutions.

The following tables are based on *Baumé's hydrometers*, a set (two) of which is assumed to be available.

The scale of the hydrometer *for liquids lighter than water* has its zero-point at the bottom. This is the point to which the instrument sinks in a 10% solution of common salt. The 10° point is that to which it sinks in pure water. One degree of the scale at any part is one tenth of this interval.

The hydrometer *for liquids heavier than water* has its zero point at the top of the scale. This is the point to which it sinks in pure water; and the 10° point is that to which it sinks in a 10% salt solution. One degree of the scale is one tenth of this interval.

A solution of a given percentage strength is prepared by adding to a number of parts of the substance to be dissolved equal to the required percentage, a sufficient quantity of the solvent to make 100 parts in all. Thus, a 10% salt solution consists of 10 parts of salt in 90 parts of water.

For practical purposes, one cubic centimeter of water or alcohol may be considered as one gram.

The figures in the following tables refer to parts by *volume* at about 60° F. Intermediate values may be obtained from those given above with sufficient accuracy by interpolation.

° Baumé.	Heavier than Water.					Lighter than Water.		
	Specific Gravity.	% KOH.	% HCl. 22° B.	% H_2SO_4.	% HNO_3.	Specific Gravity.	% Alcohol. Tralles' Scale.	% NH_3. *Italics* = % NH_4OH 26° B.
3	1.022	2.6	12.6	3.8	4.0			
5	1.037	4.5	20.4	5.8	6.3			
8	1.060	7.4	33.6	8.8	10.2			
10	1.075	9.2	42.0	10.8	12.7	1.000	0.	0.
12	1.091	10.9	50.7	13.0	15.3	.986	10.	*11.5* 3.3
15	1.116	13.8	64.7	16.2	19.4	.967	28.	*28.0* 8.0
17	1.134	15.7	74.5	18.5	22.2	.954	38.	*39.9* 11.4
20	1.162	18.6	89.6	22.2	26.3	.936	49.	*58.0* 16.6
22	1.180	20.5	100.0	24.5	29.2	.924	55.	*71.3* 20.4
25	1.210	23.3	119.0	28.4	33.8	.907	62.	*92.0* 26.3
27	1.231	25.1		31.0	37.0	.896	67.	30.7
30	1.263	28.0		34.7	41.5	.879	74.	
32	1.285	29.8		37.	45.0	.869	78.	
35	1.320	32.7		41.0	50.7	.854	83.	
37	1.345	34.9		44.4	55.0	.844	87.	
40	1.383	37.8		48.3	61.7	.829	91.	
42	1.410	39.9		51.2	67.5	.820	94.	
45	1.453	43.4		55.4	78.4	.807	97.	
47	1.483	45.8		58.3	87.1	.798	99.	
50	1.530	49.4		62.5	100.0			
66	1.842			100.0				

Table for Acetic Acid.

The difference between the specific gravities of water and of glacial acetic acid is small. The mixture of the two substances has the peculiarity that its specific gravity increases up to a certain point with the addition of acid, and then decreases to that of the pure acid. If the specific gravity is above 1.055, and is increased by the addition of water, the acid is above 78%; if it is decreased by the addition of water, the acid is below 78%.

° B.	Sp. gr.	% $HC_2H_3O_2$.	° B.	Sp. gr.	% $HC_2H_3O_2$.
1	1.0068	5.0 —	6	1.0426	31.2 —
2	1.0138	9.7 —	7	1.0501	37.9 —
3	1.0208	14.6 —	8	1.0576	45.6 or 99.1
4	1.0280	19.7 —	9	1.0653	55.0 or 95.4
5	1.0353	25.2 —	10	1.0731	69.5 or 87.0
				1.0748	77 to 80

Table for Diluting Alcohol.

Desired Strength of Alcohol.	100 volumes of alcohol of %								
	90	85	80	75	70	65	60	55	50
	require addition of vols. water								
85	6.6								
80	13.8	6.8							
75	21.9	14.5	7.2						
70	31.1	23.1	15.4	7.6					
65	41.5	33.0	24.7	16.4	8.2				
60	53.7	44.5	35.4	26.5	17.6	8.8			
55	67.9	57.9	48.1	38.3	28.6	19.0	9.5		
50	84.7	73.9	63.0	52.4	41.7	31.3	20.5	10.4	
45	105.3	93.3	81.4	69.5	57.8	46.1	34.5	22.9	11.4
40	130.8	117.3	104.0	90.8	77.6	64.5	51.4	38.5	25.6
35	163.3	148.0	132.9	117.8	102.8	87.9	73.1	58.3	43.6
30	206.2	188.6	171.1	153.6	136.4	118.9	101.7	84.5	67.5
25	266.1	245.2	224.3	203.5	182.8	162.2	141.7	121.2	100.7
20	355.8	329.8	304.0	278.3	252.6	227.0	201.4	176.0	150.6
15	505.3	471.0	436.9	402.8	268.8	334.9	301.1	267.3	233.5
10	804.5	753.7	702.9	652.2	601.6	551.1	500.6	450.2	399.9

Crystal Systems.

Name of System.	Principal Axes.		
	No.	Rel. Length.	Rel. Positions.
Cubic or monometric.	3	all equal.	at right-angles.
Tetragonal or dimetric.	3	two equal, third variable.	at right-angles.
Hexagonal	4	three equal, fourth variable.	three at 60° with each other; fourth at 90° with these.
Rhombic or trimetric.	3	of different lengths.	at right-angles.
Monoclinic or oblique.	3	of different lengths.	two at right-angles, third oblique.
Triclinic or asymmetric.	3	of different lengths.	all oblique to each other.

Isotonic Coefficient.

Two solutions of equal power to take up water are said by De Vries to be in isotonic concentration. He terms a number showing the water-absorbing power of a given solution, as compared with that of a solution of saltpeter of equal strength taken as a standard, its *isotonic coefficient*. For convenience, the coefficient assigned to saltpeter is 3.

De Vries finds that compounds fall into six groups whose coefficients are approximately whole numbers, as follows:

I. Organic compounds not containing metals, and free acids; e.g., Cane-sugar, tartaric acid, citric acid, etc. **2.**

II. Salts of alkaline earths with one acid group in the molecule; e.g., Magnesium sulphate and malate. **2.**

III. Salts of alkaline earths with two acid groups in the molecule; e.g., Magnesium chloride and citrate, calcium chloride. **4.**

IV. Salts of alkali-metals with one atom of alkali in the molecule; e.g., Potassium or sodium nitrate, chloride, acetate, etc.   3.
V. Salts of alkali-metals with two atoms of alkali in the molecule; e.g., Potassium sulphate, oxalate, tartrate, malate, etc.   4.
VI. Salts of alkali-metals with three atoms of alkali in the molecule; e.g., Potassium citrate.   5.

Thus a solution of cane-sugar has $\frac{1}{3}$ the water-attracting power of one of saltpeter of the same strength; and a solution of sugar must be $\frac{3}{1}$ as strong as one of saltpeter to produce the same osmotic effects.

LITERATURE.

ALLEN, Edwin West.—I. Untersuchungen über Holzgummi, Xylose und Xylonsäuren. Inaug.-Diss. Göttingen, 1890.

ALTMANN.—I. Die Elementarorganismen und ihre Beziehungen zu den Zellen. Leipzig, 1890.

—— II. Ueber Nucleïnsäuren. Archiv f. Anatomie u. Physiologie, Physiol. Abt. 1889. p. 524.

—— III. Die Struktur des Zellkernes. Ib. Anatom. Abt. 1889. p. 409.

AMBRONN, H.—I. Ueber das optische Verhalten der Cuticula und der verkorkten Membranen. Ber. d. D. bot. Ges. 1888. p. 226.

ARNAUD, A.—I. Recherches sur la composition de la carotine, sa fonction chimique et sa formule. Comptes rendus. 1886. T. 102. p. 1119.

—— II. Recherches sur les matières colorantes des feuilles; identité de la matière rouge orangé avec la carotine, $C_{18}H_{24}O$. Ib. 1885. T. 100. p. 751.

—— u. Padé, L.—I. Recherche chimique de l'acide nitrique, des nitrates dans les tissus végétaux. Comptes rendus. T. 98. p. 1488.

ASKENASY, E.—I. Beiträge zur Kenntniss des Chlorophylls und einiger dasselbe begleitenden Farbstoffe. Bot. Zeitg. 1867. p. 225.

BACHMANN, E.—I. Emodin in Nephroma lusitanica. Ber. d. D. bot. Ges. 1887. p. 192.

—— II. Spektroskopische Untersuchungen über Pilzfarbstoffe. Progr. d. Gymnasiums zu Plauen. Ostern 1886.

—— III. Mikrochemische Reaktionen auf Flechtenstoffe als Hilfsmittel zum Bestimmen der Flechten. Zeitschr. f. w. Mikrosk. Bd. III. p. 216.

—— IV. Ueber nicht krystallisierte Flechtenfarbstoffe. Pringsheim's Jahrbücher. Bd. XXI. p. 1.

—— V. Mikrochemische Reaktionen auf Flechtenfarbstoffe. Flora. 1887. p. 291.

—— VI. Referat über Solla (II). Zeitschrift f. w. Mikrosk. Bd. 2. p. 260.

DE BARY.—I. Ueber die Wachsüberzüge der Epidermis. Bot. Zeitg. 1871. p. 129 und 566.
—— II. Morphologie u. Biologie der Pilze. II. Aufl. Leipzig, 1884.
BEHRENS, J.—Ueber einige ätherisches Öel secernierende Hautdrüsen. Ber. d. D. bot. Ges. 1886. p. 400.
—— W.—I. Leitfaden der botanischen Mikroskopie. Braunschweig, 1890.
—— II. Tabellen zum Gebrauch bei mikroskopischen Arbeiten. Ib. 1887.
—— III. Hilfsbuch zur Ausführung mikroskopischer Untersuchungen. Ib. 1883.
BEILSTEIN.—I-III. Handbuch der organischen Chemie. II. Auflage. Leipzig, 1886–1890. Bd. I-III.
BELZUNG.—I. Recherches morphol. et physiol. sur l'amidon et les grains de chlorophylle. Ann. d. sc. nat., Bot. Sér. VII. T. 5. p. 179.
—— II. Sur divers principes issus de la germination et leur cristallisation intracellulaire. Journ. de Botanique. 1892. p. 49.
—— et Poirault.—I. Sur les sels de l'Angiopteris evecta et en particulier le malate neutre de calcium. Journ. de Botanique. 1892. p. 286.
BERTHOLD.—I. Studien über Protoplasmamechanik. Leipzig, 1886.
—— II. Beiträge zur Morphologie und Physiologie der Meeresalgen. Pringsheim's Jahrbücher. Bd. XIII. p. 569.
BOKORNY, Th.—I. Eine bemerkenswerte Wirkung oxydierter Eisenvitriollösungen auf lebende Pflanzenzellen. Ber. d. D. bot. Ges. 1889. p. 274.
—— II. Ueber den Nachweis von Wasserstoffsuperoxyd in lebenden Pflanzenzellen. Ib. p. 275.
—— III. Das Wasserstoffsuperoxyd und die Silberabscheidung durch aktives Albumin. Pringheim's Jahrb. Bd. 17. p. 347.
—— IV. Ueber Aggregation. Ib. Bd. 20. p. 427.
—— V. Ueber die Einwirkung basischer Stoffe auf das lebende Protoplasma. Ib. Bd. 19. p. 206.
—— VI. Neue Untersuchungen über den Vorgang der Silberabscheidung durch aktives Albumin. Ib. Bd. 18. p. 194.
BORODIN.—I. Ueber die mikrochemische Nachweisung und die Verbreitung des Dulcits im Pflanzenreich. Revue d. sc. n. p. p. l. Soc. d. Nat. d. St. Peters b.1890. N. 1. p. 26. (Ref.: Bot. Centralbl. 1890. Bd. 43. p. 175.)
—— II. Ueber die physiologische Rolle und die Verbreitung des Asparagins im Pflanzenreiche. Bot. Zeitung. 1878. p. 801.
—— III. Ueber einige bei Bearbeitung von Pflanzenschnitten mit Alkohol entstehende Niederschläge. Ib. 1882. p. 589.

BORODIN.—IV. Ueber Sphaerokrystalle aus Paspalum elegans und überdie mikrochemische Nachweisung von Leucin. Arbeiten d. St. Petersb. Naturf. Ges. Bd. XIII. p. 47. (Ref.: Bot. Centralbl. 1884. Bd. XVII. p. 102.)

BORSCOW, El.—I. Beiträge zur Histochemie der Pflanzen. Bot. Zeitg. 1874. p. 17.

BRAEMER, L.—Un nouveau réactiv histo-chimique des tannins. Bull. Soc. d'hist. nat. de Toulouse. Janv. 1889. (Ref.: Zeitschr. f. w. Mikroskopie. Bd. VI. p. 114.)

BREDOW, Hans.—I. Beiträge zur Kenntniss der Chromatophoren. Pringsheim's Jahrb. f. w. Bot. Bd. 22. p. 349.

BUESGEN.—I. Art und Bedeutung des Tierfanges bei Utricularia vulgaris. Ber. d. D. bot. Ges. 1888. p. LV.

BUETSCHLI, O.—I. Ueber den Bau der Bakterien und verwandter Organismen. Leipzig, 1890.

BUSSE, W.—I. Photoxylin als Einbettungsmittel für pflanzliche Objecte. Zeitsch. f. wiss. Micros. Bd. IX. p. 47. 1892.

—— II. Nachträgliche Notiz zur Celloidin-Einbettung. Ib. Bd. IX. p. 49. 1892.

CAMPBELL, Douglas H.—I. The Staining of living Nuclei. Untersuch. a. d. botan. Institut zu Tübingen. Bd. II. p. 569.

CLAUTRIAU, G.—I. Recherches microchimiques sur la localisation des alcaloides dans le Papaver somniferum. Ann. de la Soc. belge de Microsc. T. XII. 1889. p. 67. (Ref.: Zeitschr. f. w. Mikrosk. Bd. VI. p. 243.)

COHN, Ferdinand.—I. Untersuchungen über Bakterien. II. Cohn's Beitr. z. Biol. d. Pflanzen. Bd. I. Heft III. p. 141.

CORRENS, Carl Erich—I. Ueber Dickenwachstum durch Intussusception bei einigen Algenmembranen. Münchener Inaug.-Diss. 1889, u. Flora 1889.

—— II. Zur Anatomie und Entwicklungsgeschichte der extranuptialen Nectarien von Dioscorea. Sitzungsber. d. k. Akad. d. W. in Wien. Mathem.-naturw. Cl. Bd. XCVII. Abt. I. 1888. p. 652.

—— III. Zur Kenntniss der inneren Struktur der vegetabilischen Zellmembranen. Pringsheim's Jahrb. f. w. Botan. Bd. XXIII. p. 254.

COURCHET.—I. Recherches sur les chromoleucites. Annales d. sc. nat., Bot. Sér. VII. T. VII. 1888. p. 263.

CRATO, E.—I. Die Physode, ein Organ des Zellenleibes. Ber. der D. bot. Ges. Bd. X. p. 295. 1892.

CZAPLEWSKI, Eugen.—I. Die Untersuchung des Auswurfs auf Tuberkelbacillen. Jena, 1891.

DEHNECKE.—I. Ueber nicht assimilierende Chlorophyllkörper. Inaug.-Dissert. Bonn, 1880.

DETMER, W.—I. Das pflanzenphysiologische Praktikum. Jena, 1888.

DIPPEL.—I. Kalium-Quecksilberjodid als Quellungsmittel. Zeitschrift f. w. Mikroskopie. Bd. I. p. 251.

—— II. Handbuch der allgemeinen Mikroskopie. II. Aufl. Braunschweig, 1882.

DOGIEL, A. S.—I. Ein Beitrag zur Farbenfixierung von mit Methylenblau tingierten Präparaten. Zeitschr. f. w. Mikrosk. Bd. VIII. p. 15.

DRAGENDORFF, Georg.—I. Die qual. und quant. Analyse von Pflanzen und Pflanzenteilen. Göttingen, 1882.

DUFOUR, Jean.—I. Notices microchimiques sur le tissu épidermique des végétaux. Bull. de la Soc. vaud. d. Sc. nat. T. XXII. Nr. 94.

—— II. Recherches sur l'amidon soluble. Ib. T. XXI. Nr. 93.

ENGELMANN, Th. W.—I. Neue Methode zur Untersuchung der Sauerstoffausscheidung pflanzlicher und tierischer Organismen. Bot. Zeitung. 1881. p. 441.

ERRERA, Léo.—I. Ueber den Nachweis des Glycogens bei Pilzen. Bot. Zeitung. 1886. p. 316.

—— II. Sur le glycogène chez les Basidiomycètes. Mém. de l'Acad. d. Belgique. T. 37. (Ref.: Zeitschr. f. w. Mikrosk. Bd. III. p. 277.)

—— III. Sur l'emploi de l'encre de Chine en microscopie. Bull. d. l. Soc. belge de Microscopie. T. X. p. 478. (Ref.: Zeitschr. f. w. Mikrosk. Bd. II. p. 84.)

—— IV. Canarine. Ib. p. 183. (Ref.: Botan. Jahresber. 1884. p. 193.)

—— V. Sur la distinction microchimique des alcaloides et des matières protéiques. Annales de la Soc. belge de Microscopie. Mémoires. Tome XIII. (Ref.: Bot. Centralbl. 1891. Bd. 46. p. 225.)

—— Maistriau et Clautriau.—I. Premières recherches sur la localisation et la signification des alcaloides dans les plantes. Ann. de la Soc. belge de Microscopie. T. XII. 1889. p. 1. (Ref.: Zeitschr. f. w. Mikrosk. Bd. VI. p. 389.)

ETERNOD, A.—I. Instruments destinés à la microscopie. Zeitschr. f. w. Mikrosk. Bd. IV. p. 39.

EVCLESHEIMER, A. C.—I. Celloidin imbedding in plant histology. Bot. Gazette. Vol. XV. p. 272. 1890.

FISCHER, Alfred.—I. Ueber das Vorkommen von Gypskrystallen bei den Desmidiaceen. Pringsheim's Jahrb. Bd. XIV. p. 133.

—— II. Die Plasmolyse der Bakterien. Berichte der K. Sächs. Ges. d. W. Math.-phys. Kl. 1891. p. 52.

FISCHER, Alfred.—III. Ueber den Inhalt der Siebröhren in der unverletzten Pflanze. Ber. d. D. bot. Ges. 1885. p. 230.
—— IV. Neue Beobachtungen über Stärke in den Gefässen. Ib. 1886. p. XCVII.
—— V. Beiträge zur Physiologie der Holzgewächse. Pringsheim's Jahrb. Band XXII. p. 73.
—— VI. Neue Beiträge zur Kenntniss der Siebröhren. Ber. d. math.-phys. Klasse der K. Sächs. Ges. d. Wiss. 1886.
—— Emil.—I. Synthesen in der Zuckergruppe. Bericht d. D. chem. Ges. 1890. Bd. 23. p. 2114.
FLEMMING, W.—I. Ueber Teilung und Kernformen bei Leukocyten und über deren Attraktionssphären. Archiv f. mikr. Anatom. Bd. 37. p. 249.
—— II. Weiteres über die Entfärbung osmierten Fettes in Terpentin und anderen Substanzen. Zeitschrift f. w. Mikrosk. Bd. VI. p. 178.
— III. Neue Beiträge zur Kenntniss der Zelle. II. Teil. Archiv f. mikr. Anatomie. Bd. 37. p. 685.
FORSSELL, K. B. J.—I. Beiträge zur Mikrochemie der Flechten. Sitzungsber. d. K. Akad. der W. z. Wien. Mathem.-naturw. Kl. 1886. Bd. 93. Abt. I. p. 219.
FRANK, B.—I. Untersuchungen über die Ernährung der Pflanze mit Stickstoff. Landwirtsch. Jahrb. 1888. p. 421.
—— II. Bemerkungen zu vorstehendem Artikel (Kreusler I). Ib. p. 723.
GARDINER, Walter.—I. On the Phenomena accompanying Stimulation of the Gland-Cells in the Tentacles of Drosera dichotoma. Proccedings of the R. Soc. of London. V. XXXIX. p. 229.
—— II. On the continuity of the protoplasm through the walls of vegetable cells. Arb. d. bot. Instit. in Würzb. Bd. III. p. 52.
—— III. The determination of Tannin in vegetable cells. The Pharm. Journ. and Transact. 1884. N. 709. p. 588. (Ref.: Zeitschr. f. w. Mikroskop. Bd. I. p. 464.
GEOFFROY.—I. Journal de Botanique. 1893. p. 55.
GIBELLI, Giuseppe.—I. Nuovi studi sulla malattia del Castagno detta dell' inchiostro. Bologna. 1883. (Ref.: Zeitschr. f. w. Mikrosk. Bd. I. p. 137.)
GIERKE, Hans.—I. Färberei zu mikroskopischen Zwecken. Braunschweig, 1885.
GILSON, Eugène.—I. La subérine et les cellules du liège. La Cellule etc. p. p. Carnoy. T. VI. 1890. p. 63.
GILTAY, E.—I. Ueber das Verhalten von Haematoxylin gegen Pflanzenzellmembranen. Sitzungsbericht der K. Akadem. d. Wiss. zu Amsterdam. 27. Oktober 1883. p. 2.
GRAVIS, A.—I. L'Agar-Agar comme fixatif des coupes microto-

miques. Bull. d. l. Soc. belge de Microscopie. 1889. p. 72.
(Ref. : Bot. Centralblatt. 1890. Bd. 41. p. 13.)

GREEN, J. R.—I. On the germination of the tuber of the Jerusalem Artichoke. Annals of Botany. 1889. Vol. I. p. 223. (Ref.: Zeitschr. f. w. Mikrosk. 1889. Bd. 6. p. 244.)

GUENTHER, Carl.—I. Einführung in das Studium der Bakteriologie. Leipzig, 1891.

GUIGNARD, Léon.—I. Développement et constitution des anthérozoïdes. Revue gén. de Botanique. Bd. I. p. 11.

—— II. Sur la localisation des principes qui fournissent les essences sulfurées des Crucifères. Comptes rend. 1890. T. 111. p. 249.

—— III. Sur la localisation, dans les plantes, des principes qui fournissent l'acide cyanhydrique. Ib. 1890. T. 110. p. 477.

—— IV. Nouvelles études sur la fécondation. Annales d. sc. nat., Bot. Sér. VII. T. XIV. p. 163.

HABERLANDT, G.—I. Das reizleitende Gewebesystem der Sinnpflanze. Leipzig, 1890.

HANAUSEK, T. F.—I. Zur histochemischen Coffeïnreaktion. Zeitschr. d. Allg. Oesterreich. Apotheker-Vereins. 1891. p. 606. (Ref.: Bot. Centralbl. 1891. Bd. 48. p. 284.)

HANSEN, A.—I. Ueber Sphaerokrystalle. Arb. d. botan. Instituts in Würzburg. Bd. III. p. 92.

—— II. Das Chlorophyllgrün der Fucaceen. Ib. p. 288.

—— III. Die Farbstoffe der Blüten und Früchte. Verh. d. Physik.-Med. Gesellschaft zu Würzburg. N. F. Bd. XVIII. N. 7.

HANSTEIN, J.—I. Ueber eine Conferve, welche die Eigentümlichkeit hat, sich mit Gürteln von Eisenoxydhydrat zu umkleiden. Sitzungsber. d. niederrh. Ges. zu Bonn. 1878. p. 73.

HARZ, C. O.—I. Ueber Physomyces heterosporus n. sp. Botan. Centralblatt. 1890. Bd. 41. p. 405.

—— II. Ueber die Entstehung und Eigenschaften des Spergulins, eines neuen Fluorescenten. Botan. Zeitg. 1877. p. 489.

HAUG, R.—I. Winke zur Darstellung von Präparaten von intra vitam mit Anilinfarben injizierten Geschwulstpartien. Zeitschr. f. w. Mikrosk. Bd. VIII. p. 11.

HAUPTFLEISCH, Paul.—I. Zellmembran und Hüllgallerte der Desmidiaceen. Inaug.-Dissert. Greifswald, 1888.

HAUSHOFER, H.—I. Mikroskopische Reaktionen. Braunschweig, 1885.

HEGLER, Robert.—I. Histochemische Untersuchungen verholzter Zellmembranen. Flora. 1890. p. 31.

HEINE, H.—I. Ueber die physiologische Funktion der Stärkescheide. Ber. d. D. bot. Ges. 1885. p. 189.

HEINRICHER, E.—I. Verwendbarkeit der Eau de Javelle zum Nach-

weis kleiner Stärkemengen. Zeitschr. f. w. Mikrosk. Bd. III. p. 213.
HENKING.—I. Ein einfaches Mikrotommesser. Zeitschr. f. w. Mikroskopie. 1885. Bd. II. p. 509.
HERMANN, F.—I. Beiträge zur Histologie des Hodens. Archiv f. mikrosk. Anatomie. Bd. 34. p. 58.
—— II. Beiträge zur Lehre von der Entstehung der karyokinetischen Spindel. Ib. Bd. 37. p. 569.
HERRMANN, Ottomar.—I. Nachweis einiger organischer Verbindungen in den vegetabilischen Geweben. Inaug.-Diss. Leipzig, 1876.
HERTWIG, O.—I. Die Zelle und die Gewebe. Jena, 1892.
HIERONYMUS, G.—I. Ueber Dicranochaete reniformis Hieron., eine neue Protococcacea des Süsswassers. Cohn's Beiträge z. Biolog. der Pflanzen. Bd. V. p. 351.
—— II. Beiträge zur Morphologie und Biologie der Algen. II. Cohn's Beiträge zur Biol. d. Pfl. Bd. V. p. 461. 1892.
v. HÖHNEL.—I. Ueber den Kork und verkorkte Gewebe überhaupt. Sitzungsber. d. Akad. der Wiss. zu Wien. Bd. 76, I. p. 507.
—— II. Histochemische Untersuchungen über das Xylophilin und das Coniferin. Ib. Band 67, I. p. 663.
HOFFMEISTER, W.—I. Die Rohfaser und einige Formen der Cellulose. Landwirtschaftl. Jahrbücher. 1888. p. 239.
—— II. Die Cellulose und ihre Formen. Ib. 1889. p. 767.
HOLZNER, Georg.—I. Ueber Krystalle in den Pflanzenzellen. Inaug.-Dissert. und Flora. 1864.
HUEPPE, Ferdinand.—I. Die Methoden der Bakterien-Forschung. V. Auflage. Wiesbaden, 1891.
HUSEMANN, A. und A. Hilger und Th. Husemann.—I. Die Pflanzenstoffe. II. Aufl. 1882–84.
IHL, Anton.—I. Einwirkung der Phenole auf Cinnamaldehyd. Zimmtaldehyd, ein wahrscheinlicher Bestandteil der Holzsubstanz. Chemikerzeitung. 1889. p. 560. (Ref.: Bot. Centralblatt. 1889. Bd. 39. p. 184.)
—— II. Ueber neue empfindliche Holzstoff- und Cellulose-Reagentien. Ib. 1885. p. 266. (Ref.: Zeitschr. für w. Mikrosk. Bd. II. p. 259.)
IMMENDORFF, H.—I. Das Carotin im Pflanzenkörper und Einiges über den grünen Farbstoff des Chlorophyllkorns. Landwirtschaftl. Jahrbücher. 1889. p. 507.
JÖNSSON, B.—I. Entstehung schwefelhaltiger Oelkörper in den Mycelfäden von Penicillium glaucum. Botan. Centralbl. 1889. Bd. 37. p. 201.
KAERNER, W.—I. Ueber den Abbruch und Abfall pflanzlicher Behaarung und den Nachweis von Kieselsäure in Pflanzenhaaren.

Nova Acta d. Ksl. Leop.-Carol. D. Acad. d. Naturf. 1889.
Bd. 54. N. 3. p. 219.

KIENITZ-GERLOFF.—I. Die Protoplasmaverbindungen zwischen benachbarten Gewebselementen in der Pflanze. Botan. Zeitg. 1891. p. 1.

KIRCHNER.—I. Die mikroskopische Pflanzenwelt des Süsswassers. Braunschweig, 1885.

KLEBAHN.—I. Studien über Zygoten I. Pringsheim's Jahrb. Bd. 22. p. 415.

KLEBS.—I. Ueber die Organisation einiger Flagellatengruppen. Untersuch. a. d. bot. Institut zu Tübingen. Bd. I. p. 233.

—— II. Ueber die Organisation der Gallerte bei einigen Algen und Flagellaten. Ib. Bd. II. p. 333.

—— III. Beiträge zur Physiologie der Pflanzenzelle. Ib. p. 489.

—— IV. Einige Bemerkungen zu der Arbeit von Krasser " Untersuchungen über das Vorkommen von Eiweiss in der pflanzlichen Zellhaut, etc." Botan. Zeitg. 1887. p. 697.

—— V. Ein kleiner Beitrag zur Kenntniss der Peridineen. Ib. 1884. p. 721.

KLEIN, J.—I. Die Krystalloide der Meeresalgen. Pringsheim's Jahrb. Bd. XIII. p. 23.

KLERCKER, John af.—I. Studien über die Gerbstoffvakuolen. Tübinger Inaugur.-Dissert. 1888.

—— II. Ueber das Kultivieren lebender Organismen unter dem Mikroskop. Zeitschrift f. w. Mikrosk. Bd. VI. p. 145.

—— III. Ueber Dauerpräparate gerbstoffhaltiger Objecte. Verh. d. biol. Vereins in Stockholm. Bd. VI. No. 3. 1891.

KOCH, L.—I. Microtechnische Mittheilungen. Pringsheim's Jahrbücher. Bd. XXIV. p. 1. 1892.

KOHL, F. G.—I. Protoplasmaverbindungen bei Algen. Bericht d. D. botan. Ges. 1891. p. 9.

—— II. Anatomisch-physiologische Untersuchung der Kalksalze und Kieselsäure in der Pflanze. Marburg, 1889.

KRABBE, G.—I. Untersuchungen über das Diastaseferment unter spezieller Berücksichtigung seiner Wirkung auf Stärkekörner innerhalb der Pflanze. Pringsheim's Jahrb. f. w. Bot. Bd. 21. p. 520.

KRASSER, Fridolin.—I. Untersuchungen über das Vorkommen von Eiweiss in der pflanzlichen Zellhaut, nebst Bemerkungen über den mikrochemischen Nachweis der Eiweisskörper. Sitzungsbericht der K. Akad. der Wiss. zu Wien. 1886. Bd. 94. Abt. I. p. 118.

—— II. Ueber eine Conservirungsflüssigkeit und die fixirende Eigenschaft des Salicylaldehyds. Bot. Centralbl. Bd. LII. p. 4. 1892.

KRASSER FRIDOLIN.—III. Ueber neue Methoden zur dauerhaften Präparation des Aleuron und seiner Einschlüsse. Sitzungsber. d. zool.-bot. Ges. zu Wien. Bd. XLI. 1891.
KREUSLER.—I. Zum Nachweis von Nitraten im Erdboden, etc. Landwirtsch. Jahrbücher. 1888. p. 721.
KUGLER, Karl.—I. Ueber das Suberin. Strassburger Inaug.-Dissert. 1884.
LAGERHEIM, G.—I. Ueber die Anwendung von Milchsäure bei der Untersuchung trockener Algen. Hedwigia. 1888. p. 58. (Ref.: Ztschr. f. w. Mikr. Bd. V. p. 552.)
—— II. L'acide lactique, excellent agent pour l'étude des champignons secs. Revue mycologique. T. XI. 1889. p. 95. (Ref.: Ib. Bd. VI. p. 380.)
LANGE, Gerhard.—I. Zur Kenntniss des Lignins. I. Zeitschrift für physiologische Chemie. Bd. XIV. p. 15.
—— II. Id. II. Mitteilung. Ib. p. 217.
—— Theodor.—I. Beiträge zur Kenntniss der Entwicklung der Gefässe und Tracheïden. Flora. 1891. p. 393.
LECOMTE, Henri.—I. Contribution à l'étude du liber des Angiospermes. Ann. des sc. nat., Bot. Sér. 7. T. 10. p. 193.
LEITGEB, H.—I. Ueber die durch Alkohol in Dahliaknollen hervorgerufenen Ausscheidungen. Botan. Zeitung. 1887. p. 129.
—— II. Der Gehalt der Dahliaknollen an Asparagin und Tyrosin. Mitteilungen a. d. Botan. Instit. zu Graz. Heft II. p. 215.
LINDT, O.—I. Ueber den Nachweis von Phloroglucin. Zeitschr. f. w. Mikroskop. Bd. 2. p. 495.
—— II. Ueber den mikrochemischen Nachweis von Brucin und Strychnin. Ib. Bd. I. p. 237.
LOEFFLER.—I. Weitere Untersuchungen über die Beizung und Färbung der Geisseln bei den Bakterien. Centralbl. für Bakteriol. u. Parasit. 1890. Bd. VII. p. 625.
LOEW, O.—I. Noch einmal über das Protoplasma. Botan. Zeitg. 1884. p. 113.
—— II. Ueber den mikrochemischen Nachweis von Eiweissstoffen. Ib. p. 273.
—— und Bokorny.—I. Ueber das Verhalten der Pflanzenzellen zu stark verdünnter alkalischer Silberlösung. II. Botan. Centralblatt. 1889. Bd. 39. p. 369.
—— II. Die chemische Kraftquelle im lebenden Protoplasma. München, 1882.
LUDTKE, Franz.—I. Beiträge zur Kenntniss der Aleuronkörner. Pringsheim's Jahrb. Bd. 21. p. 62.
LUNDSTRÖM, Axel N.—I. Ueber farblose Oelplastiden und die biologische Bedeutung der Oeltropfen gewisser Potamogeton-Arten. Botan. Centralblatt. Bd. 35. p. 177.

MALFATTI, H.—I. Zur Chemie des Zellkerns. Ber. der naturw.-med. Vereins in Innsbruck. XX. Jahrg. 1891-2. (Ref.: Bot. Centralbl. LV. 152.)

—— II. Beiträge zur Kenntniss der Nucleïne. Zeitsch. für physiol. Chemie. Bd. XVI. p. 68. 1892. (Ref.: Bot. Centralbl. LV. 154.)

MANGIN, Louis.—I. Observations sur la membrane du grain de pollen mur. Bull. d. l. soc. bot. de France. T. 36. 1889. p. 274.

—— II. Sur la callose, nouvelle substance fondamentale existant dans la membrane. Comptes rendus. T. 110. 1890. p. 644.

—— III. Sur la structure des Péronosporées. Ib. T. 111. 1890. p. 923.

—— IV. Sur la présence des composés pectiques dans les végétaux. Ib. T. 109. 1889. p. 579.

—— V. Sur les réactifs colorants des substances fondamentales de la membrane. Ib. 1890. T. 111. p. 120. (Ref.: Zeitschr. f. w. Mikroskopie. Bd. VII. p. 409.)

—— VI. Sur la substance intercellulaire. Ib. T. 110. p. 295. (Ref.: Ib. p. 545.)

—— VII. Sur les réactifs jodés de la cellulose. Bull. d. l. soc. bot. d. France. T. 35. 1888. p. 421. (Ref.: Ib. Bd. VI. p. 242.)

—— VIII. Observations sur la membrane cellulosique. Comptes rendus. T. CXIII. p. 1069. 1891.

—— IX. Sur la constitution des cystoliths et des membranes incrustées de carbonate de chaux. Ib. T. CXV. p. 260. 1892.

MATTIROLO, O.—I. Skatol e Carbazol, due nuovi reagenti per le membrane lignificate. Zeitschr. f. w. Mikrosk. Bd. II. p. 354.

MAYER, P.—I. Aus der Mikrotechnik. Internat. Monatschr. f. Anatom. u. Physiol. Bd. IV. 1887. (Ref.: Zeitschr. f. w. Mikrosk. Bd. IV. p. 76.)

—— II. Einfache Methode zum Aufkleben mikroskopischer Schnitte. Mitt. a. d. Zool. Stat. Neapel. Bd. IV. 1883. p. 521. (Ref.: Ib. Bd. II. p. 225.)

—— III. Ueber das Farben mit Carmin, Cochenille und Haematein Thonerde. Ib. Bd. X. p. 480. 1892.

MELNIKOFF.—I. Untersuchungen über das Vorkommen des kohlensauren Kalkes in Pflanzen. Inaug.-Diss. Bonn, 1877.

MESNARD, E.—I. Recherches sur la mode de production de parfum dans les fleurs. Comptes rendus. T. CXV. p. 892. 1892. (Ref.: Bot. Zeitung. LI. 185.)

MEYER, Arthur.—I. Ueber Stärkekörner, welche sich mit Jod rot färben. Berichte d. D. bot. Ges. 1886. p. 337.

Meyer, Arthur.—II. Das Chlorophyllkorn. Leipzig. A. Felix, 1883.
(Ref. : Zeitschr. f. w. Mikrosk. Bd. I. p. 302.)
—— III. Kritik der Ansichten von Frank Schwarz über die alkalische Reaktion des Protoplasmas. Bot. Zeitg. 1890. p. 234.
—— IV. Mikrochemische Reaktion zum Nachweis der reduzierenden Zuckerarten. Ber. d. D. botan. Ges. 1885. p. 332.
—— V. Chloralkarmin zur Färbung der Zellkerne der Pollenkörner. Ib. Bd. X. p. 363. 1892.
Migula.—I. Beiträge zur Kenntniss des Gonium pectorale. Botan. Centralbl. 1890. Bd. 44. p. 72.
Miliarakis.—I. Die Verkieselung lebender Elementarorgane bei den Pflanzen. Würzburg 1884. Inaug.-Diss.
Moeller, Hermann.—I. Anatomische Untersuchungen über das Vorkommen der Gerbsäure. Ber. d. D. bot. Ges. 1888. p. LXVI.
—— II. Ueber eine neue Methode der Sporenfärbung. Centralblatt f. Bakteriologie und Parasitenkunde. 1891. Bd. X. p. 273.
Molisch.—I. Grundriss einer Histochemie der pflanzlichen Genussmittel. Jena, 1891.
—— II. Ueber merkwürdig geformte Proteïnkörper in den Zweigen von Epiphyllum. Ber. d. D. botan. Gesellsch. 1885. p. 195.
—— III. Ein neues Coniferinreagenz. Ib. 1886. p. 301.
—— IV. Zur Kenntniss der Thyllen, nebst Beobachtungen über Wundheilung in der Pflanze. Sitzungsber. d. K. Akad. d. W. in Wien. Math.-nat. Kl. Bd. 97. Abt. I. p. 264.
—— V. Zwei neue Zuckerreaktionen. Ib. 1886. Bd. 93. Abt. II. p. 912.
—— VI. Ueber den mikrochemischen Nachweis von Nitraten und Nitriten in der Pflanze mittelst Diphenylamin oder Brucin. Bericht der D. botan. Gesellsch. 1883. p. 150.
—— VII. Die Pflanze in ihren Beziehungen zum Eisen. Jena, 1892.
—— VIII. Bemerkungen über den Nachweis von maskirten Eisen. Ber. der D. bot. Ges. Bd. XI. p. 73. 1893.
Moll, J. W.—I. Eene nieuwe mikrochemische looizuurreactie. Maandblad voor Natuurwetenschappen. 1884. (Ref. : Bot. Centralbl. 1885. Bd. 24. p. 250.)
Monteverde.—I. Ueber die Ablagerung von Calcium- und Magnesium-Oxalat in der Pflanze. Petersburg, 1889. (Ref. : Bot. Centralbl. 1890. Bd. 43. p. 327.)
—— II. Ueber Krystallablagerungen bei den Marattiaceen. Arb. d. St. Petersb. Naturf. Ges. 1886. Bd. 17. p. 33. (Ref.: Ib. Bd. 39. p. 358.
Mueller, Carl Oscar.—I. Ein Beitrag zur Kenntniss der Eiweissbildung in der Pflanze. Leipziger Inaug.-Diss. 1886. (Sep.-Abdr. aus Landw. Versuchsstat.)

MUELLER, C. O.—II. Kritische Untersuchungen über den Nachweis maskirten Eisens, etc. Ber. der D. bot. Gesell. Bd. XI. p. 252. 1893.
—— N. J. C.—I. Spectralanalyse der Blütenfarben. Pringsheim's Jahrbücher. Bd. XX. p. 78.
NADELMANN, Hugo.—I. Ueber die Schleimendosperme der Leguminosen. Pringsheim's Jahrbücher. Bd. 21. p. 609.
NAEGELI, C. und S. Schwendener.—I. Das Mikroskop. II. Aufl. 1877.
—— W.—I. Beiträge zur näheren Kenntniss der Stärkegruppe. Münchener Inaug.-Dissert. Leipzig, 1874.
NEBELUNG, Hans.—I. Spektroskopische Untersuchungen der Farbstoffe einiger Süsswasseralgen. Bot. Zeitg. 1878. p. 369.
NICKEL, Emil.—I. Die Farbenreaktionen der Kohlenstoffverbindungen. II. Aufl. Berlin, 1890.
—— II. Bemerkungen über die Farbenreaktionen und die Aldehydnatur des Holzes. Botan. Centralblatt. 1889. Bd. 38. p. 753.
NIGGL.—I. Das Indol ein Reagenz auf verholzte Membranen. Flora 1881. p. 545.
NOBBE, Hänlein und Councler.—I. Vorl. Notiz betr. d. Vorkommen von phosphorsaurem Kalk in der lebenden Pflanzenzelle. Landw. Versuchsstat. 1879. Bd. 23. p. 471.
NOLL, Fritz.—I. Experimentelle Untersuchungen über das Wachstum der Zellmembranen. Würzburger Habilitationsschrift u. Abhandl. der Senckenberg. naturf. Gesellsch. Bd. XV. 1887. p. 101.
OBERSTEINER, H.—I. Ein Schnittsucher. Zeitschr. f. w. Mikrosk. Bd. III. p. 55.
OVERTON.—I. Mikrotechnische Mitteilungen. Zeitschrift f. w. Mikroskopie. 1890. Bd. VII. p. 9.
—— II. Beiträge zur Histologie und Physiologie der Characeen. Botan. Centralbl. 1890. Bd. 44. p. 1.
PALLA, Ed.—I. Beobachtungen über Zellhautbildung an des Zellkernes beraubten Protoplasten. Flora. 1890. p. 314.
PFEFFER.—I. Untersuchungen über die Proteïnkörner und die Bedeutung des Asparagins beim Keimen der Samen. Pringsheim's Jahrb. Bd. VIII. p. 429.
—— II. Ueber Aufnahme von Anilinfarben in lebenden Zellen. Untersuchungen a. d. botan. Instit. zu Tübingen. Bd. II. p. 179.
—— III. Hesperidin, ein Bestandteil einiger Hesperideen. Botan. Zeitung. 1874. p. 529.
—— IV. Beiträge zur Kenntniss der Oxydationsvorgänge in lebenden Zellen. Abhandlungen der mathem.-phys. Kl. d. K. Sächs. Gesellsch. d. W. Bd. XV. p. 375.

PFEFFER.—V. Osmotische Untersuchungen. Leipzig, 1877.
—— VI. Die Oelkörper der Lebermoose. Flora. 1874. p. 2.
—— VII. Löw und Bokorny's Silberreduktion in Pflanzenzellen. Ib. 1889. p. 46.
—— VIII. Zur Kenntniss der Plasmahaut und der Vakuolen. Abhandl. d. math.-phys. Kl. der Kgl. Sächs. Ges. d. Wiss. Bd. XVI. p. 187.
—— IX. Studien zur Energetik der Pflanzen. Ib. Bd. XVIII. p. 151. 1892.
PFEIFFER, Ferdinand, R. v. Wellheim.—I. Mitteilungen über die Anwendbarkeit des venetianischen Terpentins bei botanischen Dauerpräparaten. Zeitschrift f. w. Mikrosk. Bd. 8. p. 29.
PFITZER, E.—I. Ueber ein Härtung und Färbung vereinigendes Verfahren für die Untersuchung des plasmatischen Zellleibes. Berichte d. D. botan. Gesellsch. 1883. p. 44.
PLUGGE.—I. Salpetrige Säure-haltiges Quecksilbernitrat als Reagenz auf aromatische Körper mit einer Gruppe OH am Benzolkern. Archiv der Pharmacie. 1890. Bd. 228. p. 9.
POULSEN.—I. Botanische Mikrochemie. Uebersetzt von C. Müller. Cassel, 1881. Engl. trans. by Trelease. Boston, 1882.
—— II. Note sur la préparation des grains d'aleurone. Revue gén. d. Botan. 1890. p. 547.
PRAEL, Edmund.—I. Vergleichende Untersuchungen über Schutz- und Kern-holz der Bäume. Pringsheim's Jahrb. Bd. XIX. p. 1.
PRINGSHEIM.—I. Ueber Cellulinkörner. Ber. d. D. botan. Ges. 1883. p. 288.
—— II. Ueber Lichtwirkung und Chlorophyllfunktion. Pringsheim's Jahrbücher. Bd. 12. p. 288.
RABL.—I. Ueber Zellteilung. Morphologisches Jahrbuch. Bd. X. 1885. p. 214.
RADLKOFER, L.—I. Zur Klärung von Theophrasta und der Theophrasteen. Sitzungsbericht der math.-physik. Kl. d. k. B. Akad. d. Wiss. zu München. 1889. Bd. 19. p. 221.
—— II. Ueber die Gliederung der Familie der Sapindaceen. Ib. 1890. Bd. 20. p. 105.
RANVIER.—I. Technisches Lehrbuch der Histologie. Uebersetzt v. Nicati und von Wyss. Leipzig, 1888.
REICHL, C. und C. Mikosch.—I. Ueber Eiweissreaktionen und deren mikrochemische Anwendung. Jahresbericht der K. K. Oberrealschule in d. II. Bezirke von Wien. Wien, 1890.
REINITZER, Friedrich.—I. Bemerkungen zur Physiologie des Gerbstoffs. Ber. d. D. botan. Ges. 1889. p. 187.
—— II. Ueber die wahre Natur des Gummifermentes. Zeitschrift f. physiol. Chemie. Bd. 14. p. 453.

REINKE, Friedrich.—I. Untersuchungen über das Verhältniss der von Arnold beschriebenen Kernformen zur Mitose u. Amitose. Inaug.-Diss. Kiel, 1891.
—— J.—I. Die chemische Zusammensetzung des Protoplasma von Aethalium septicum. Untersuch. a. d. botan. Labor. d. Univ. Göttingen. Heft II. p. 1.
—— II. Beitrag zur Kenntniss des Phycoxanthins. Pringsheim's Jahrbücher. Bd. X. p. 399.
REISS.—I. Ueber die Natur der Reservecellulose und über ihre Auflösungsweise bei der Keimung der Samen. Landwirtschaftliche Jahrbücher. Bd. XVIII. 1889. p. 711.
RHUMBLER, L.—I. Die verschiedenen Cystenbildungen und die Entwicklungsgeschichte der holotrichen Infusoriengattung Colpoda. Zeitschr. f. wiss. Zoologie. Bd. 46. 1888. p. 549. (Ref.: Zeitschr. f. w. Mikrosk. Bd. VI. p. 50.)
RICHTER, K.—I. Beiträge zur genaueren Kenntniss der chemischen Beschaffenheit der Zellmembran bei den Pilzen. Sitzungsber. d. Akadem. d. Wiss. zu Wien. Bd. 83. I. p. 494.
ROSEN, F.—I. Beiträge zur Kenntniss der Pflanzenzellen. Cohn's Beitr. zur Biol. d. Pfl. Bd. V. p. 443. 1892.
ROSOLL, Alexander.—I. Ueber den mikrochemischen Nachweis der Glykoside und Alkaloide in den vegetabilischen Geweben. 25. Jahresber. des Landes-Realgymnasiums zu Stockerau. 1889-90.
—— II. Beiträge zur Histochemie der Pflanzen. Sitzungsbericht d. Akad. d. Wiss. zu Wien. 1884. Bd. 89. Abt. I. Mathem.-naturw. Kl.
ROSTAFINSKI, J.—Ueber den roten Farbstoff einiger Chlorophyceen, sein sonstiges Vorkommen und seine Verwandtschaft zum Chlorophyll. Botan. Zeitung. 1881. p. 461.
RUSSOW.—I. Ueber die Verbreitung der Callusplatten bei den Gefässpflanzen. Sitzungsbericht d. naturf. Gesellsch. d. Univ. Dorpat. Bd. 6. p. 63.
SACHS, Julius v.—I. Ueber die Stoffe, welche das Material zum Wachstum der Zellhäute liefern. Pringsheim's Jahrb. Bd. III. p. 183.
SANIO, Carl.—I. Ueber die in der Rinde dicot. Holzgewächse vorkommenden krystallinischen Niederschläge und deren anatomische Verbreitung. Monatsber. d. Berl. Akad. 1857. p. 252.
—— II. Einige Bemerkungen über den Bau des Holzes. Botan. Zeitung. 1860. p. 193.
SCHENCK, H.—I. Ueber Konservierung von Kernteilungsfiguren. Inaug.-Diss. Bonn, 1890. (Ref.: Zeitschr. f. w. Mikrosk. Bd. VII. p. 38.)

SCHIMPER.—I. Ueber die Krystallisation der eiweissartigen Substanzen. Zeitschrift f. Krystallogr. u. Mineral. Bd. V. 1881. p. 131.
—— II. Zur Frage der Assimilation der Mineralsalze durch die grüne Pflanze. Flora. 1890. p. 207-261.
—— III. Untersuchungen über die Chlorophyllkörper und die ihnen homologen Gebilde. Pringsheim's Jahrb. Bd. 16. p. 1.
SCHÖNFELD, Selmar.—I. Modification of Pagan's growing slide. Journ. R. Microsc. Soc. 1888. pt. 6. p. 1028. (Ref.: Zeitschr. f. w. Mikroskopie. Bd. VI. p. 51).
SCHOTTLÄNDER, P.—I. Beiträge zur Kenntniss des Zellkerns und der Sexualzellen bei Kryptogamen. Cohn's Beiträge zur Biol. d. Pfl. Bd. VI. p. 267. 1892.
SCHMITZ.—I. Die Chromatophoren der Algen. Bonn, 1882.
—— II. Beiträge zur Kenntniss der Chromatophoren. Pringsheim's Jahrbücher f. w. Bot. Bd. XV. p. 1.
SCHUTT, Franz.—I. Ueber Peridineenfarbstoffe. Ber. d. D. bot. Ges. 1890. p. 9.
—— II. Ueber das Phycoërythrin. Ib. 1888. p. 36.
—— III. Ueber das Phycophaein. Ib. 1887. p. 259.
—— IV. Weitere Beiträge zur Kenntniss des Phycoërythrins. Ib. 1888. p. 305.
SCHULZE, E.—I. Ueber die stickstofffreien Reservestoffe einiger Leguminosensamen. Ber. d. D. botan. Ges. 1889. p. 355.
—— II. Zur Chemie der pflanzlichen Zellmembranen. II. Abhandlung. Zeitsch. für physiol. Chemie. Bd. XVI. p. 387. 1892. (Ref.: Bot. Centralbl. LV. 157.)
—— E., E. Steiger und W. Maxwell.—I. Zur Chemie der Pflanzenzellmembranen. I. Abhandlung. Zeitschrift f. physiologische Chemie. Bd. 14. 1890. p. 227.
—— Franz Eilhard.—I. Ein Entwässerungsapparat. Archiv f. mikrosk. Anatomie. Bd. 26. p. 539.
SCHWARZ, Frank.—I. Die morphologische und chemische Zusammensetzung des Protoplasmas. Breslau, 1887. Cohn's Beiträge zur Biologie der Pflanzen. Bd. V. H. 1.
—— II. Chemisch-botanische Studien über die in den Flechten vorkommenden Flechtensäuren. Ib. Bd. III. p. 249.
SHIMOYAMA.—I. Beiträge zur Kenntniss des japanischen Klebreisses. Inaug.-Diss. Strassburg, 1886.
SINGER.—I. Beiträge zur näheren Kenntniss der Holzsubstanz und der verholzten Gewebe. Sitzungsber. d. Wiener Akad. d. W. Bd. 85. Abt. I. p. 345.
SOLEREDER, H.—I. Studien über die Tribus der Gaertnereen Benth.-Hook. Ber. d. D. bot. Gesellsch. 1890. p. (71).
SOLLA, R. F.—I. Zur näheren Kenntniss der chemischen und physi-

kalischen Beschaffenheit der Intercellularsubstanz. Oesterr. bot. Zeitschr. 1879, November. (Ref.: Bot. Jahresber. 1880. p. 8.)

SOLLA. R. F.—II. Ueber zwei wahrscheinliche mikrochemische Reaktionen auf Schwefelcyanallyl. Botan. Centralbl. 1884. Bd. XX. p. 342.

SPATZIER, W.—I. Pringsh. Jahrbücher. Bd. XXV. p. 39. 1893.

STEINACH, Eugen.—I. Siebdosen, eine Vorrichtung zur Behandlung mikroskopischer Präparate. Zeitschrift f. w. Mikrosk. Bd. IV. p. 433.

STRASBURGER.—I. Das botanische Praktikum. II. Auflage. 1887.

STRENG, A.—I. Ueber eine neue mikroskopisch-chemische Reaktion auf Natrium. 24 Ber. d. Oberh. Gesellsch. f. Nat. u. Heilk. Giessen, 1885. p. 56. (Ref.: Zeitschr. f. w. Mikrosk. Bd. III. p. 129.)

—— II. Ueber einige mikroskopisch-chemische Reaktionen. Ib. p. 54. (Ref.: Ib. Bd. II. p. 429.)

—— III. Ueber einige mikroskopisch-chemische Reaktionen. Neues Jahrbuch für Mineralogie. 1888. Bd. II. p. 142. (Ref.: Zeitschr. f. w. Mikroskopie. Bd. V. p. 554.)

SUCHANNEK.—I. Technische Notiz betreffend die Verwendung des Anilinöls in der Mikroskopie sowie einige Bemerkungen zur Paraffineinbettung. Zeitschrift f. w. Mikrosk. Bd. VII. p. 156.

TEMME.—I. Ueber Schutz- und Kernholz, seine Bildung und physiologische Bedeutung. Landwirtsch. Jahrb. 1883. p. 173.

TERLETZKI, P.—I. Anatomie der Vegetationsorgane von Struthiopteris germanica Willd und Pteris aquilina L. Pringsheim's Jahrb. Bd. 15. p. 452.

THÖRNER, W.—I. Ueber den im Agaricus atrotomentosus vorkommenden chinonartigen Körper. Bericht d. D. chemisch. Gesellsch. 1878. p. 533 und 1879. p. 1630.

VAN TIEGHEM.—I. Sur les globules amylacés des Floridées. Comptes rendus. T. 61. 1865. p. 804.

—— et Douliot.—I. Recherches comparatives sur l'origine des membres endogènes. Ann. des sc. nat., Bot. Sér. VII. T. 8.

TRENKMANN.—I. Die Färbung der Geisseln von Spirillen und Bacillen. Centralblatt f. Bacteriol. und Parasitenk. 1889. Bd. VI. p. 433.

TREUB.—I. Quelques Recherches sur le rôle du noyau dans la division des cellules végétales. Naturk. Verh. d. K. Akad. Vol. XIX. Amsterdam, 1878.

VINASSA, E.—I-III. Beiträge zur pharmakognostischen Mikroskopie. Zeitschrift f. w. Mikroskopie. Bd. II. p. 309. Bd. IV. p. 295. u. Bd. VIII. p. 34.

VIRCHOW, H.—Ueber die Einwirkung des Lichtes auf Gemische von chromsauren Salzen (resp. Chromsäure), Alkohol und extra hierten organischen Substanzen. Archiv f. mikrosk. Anatomie. 1885. Bd. XXIV. p. 117. (Ref.: Zeitschr. f. w. Mikrosk. Bd. II. p. 272.)

VOIGT, A.—I. Lokalisierung des ätherischen Oeles in den Geweben der Allium-Arten. Jahrbuch der Hamburgischen wissenschaftlichen Anstalten, VI. 1889. (Ref.: Bot. Centralbl. 1890. Bd. 41. p. 292.)

VOSSELER.—I. Einige Winke für die Herstellung von Dauerpräparaten. Zeitschr. f. w. Mikrosk. 1890. Bd. VII. p. 457.

—— Venetianisches Terpentin als Einschlussmittel für Dauerpräparate. Ib. Bd. VI. p. 292.

DE VRIES,—I. Plasmolytische Studien über die Wand der Vakuolen. Pringsheim's Jahrb. Bd. 16. p. 465.

—— II. Ueber die Aggregation im Protoplasma von Drosera rotundifolia. Botan. Zeitung. 1886. p. 1.

WAAGE, Th.—I. Ueber das Vorkommen und die Rolle des Phloroglucins in der Pflanze. Ber. d. D. botan. Ges. 1890. p. 250.

WAKKER, J. H.—I. Studien über die Inhaltskörper der Pflanzenzellen. Pringsheim's Jahrbücher. Bd. 19. p. 423.

WEBER VAN BOSSE, A.—I. Études sur les Algues de l'Archipel Malaisien. II. Annales Jard. Bot. de Buitenzorg. T. VIII. p. 165. 1892.

WEHMER.—I. Das Calciumoxalat der oberirdischen Teile von Crataegus Oxyacantha L. im Herbst und im Frühjahr. Ber. d. D. botan. Gesellsch. 1889. p. 216.

WEISS, Adolf J. und Julius Wiesner.—I. Vorläufige Notiz über die direkte Nachweisung des Eisens in den Zellen der Pflanzen. Sitzungsber. d. Wiener Ak. der Wiss. Math.-naturw. Kl. 1860. Bd. XL. p. 276.

WENT, F. A. F. C.—I. Die Vermehrung der Vakuolen durch Teilung. Pringsheim's Jahrbücher. Bd. XIX. p. 295.

DE WÈVRE, A.—I. Localisation de l'atropine. Bull. Soc. belge de Microsc. T. XIII. 1887. p. 19. (Ref.: Zeitschr. f. w. Mikrosk. Bd. V. p. 119.)

WIESNER.—I. Beobachtungen über die Wachsüberzüge der Epidermis. Botan. Zeitung. 1871. p. 769.

—— II. Ueber die krystallinische Beschaffenheit der geformten Wachsüberzüge pflanzlicher Oberhäute. Ib. 1876. p. 225.

—— III. Note über das Verhalten des Phloroglucins und einiger verwandter Körper zur verholzten Zellmembran. Sitzungsber. d. Akad. d. W. zu Wien. Math.-naturw. Kl. Bd. 77. Abt. 1. 1878. p. 60.

—— IV. Ueber das Gummiferment, ein neues diastatisches Enzym,

welches die Gummi- und Schleimmetamorphose in der Pflanze bedingt. Ib. 1885. Bd. 92. Abt. I. p. 41.
WIESNER.—V. Untersuchungen über die Organisation der vegetabilischen Zellhaut. Ib. 1886. Bd. 93. Abt. I. p. 17.
WINOGRADSKY, Sergius.—I. Ueber Schwefelbakterien. Bot. Zeitg. 1887. p. 489.
—— II. Ueber Eisenbakterien. Ib. 1888. p. 261.
WINTERSTEIN, E.—I. Ueber das pflanzliche Amyloid. Zeitsch. für physiol. Chemie. Bd. XVII. p. 353. 1892. (Ref.: Bot. Centralbl. LV. 149.)
VAN WISSELINGH, C.—I. Sur la lamelle subéreuse et la subérine. Archiv néerland. T. XXVI. p. 305. 1893. (Ref.: Bot. Centralbl. LV. 109.)
WOTHTSCHALL, E.—Ueber die mikrochemischen Reaktionen des Solanin. Zeitschr. f. w. Mikrosk. Bd. V. p. 19.
ZACHARIAS, E.—I. Ueber Eiweiss, Nukleïn und Plastin. Botan. Zeitung. 1883. p. 209.
—— II. Beiträge zur Kenntniss des Zellkernes und der Sexualzellen. Ib. 1887. Nr. 18 bis 24.
—— III. Ueber die Zellen der Cyanophyceen. Ib. 1890. Nr. 1–5.
—— IV. Ueber das Wachstum der Zellhaut bei Wurzelhaaren. Flora. 1891. p. 467.
—— V. Ueber die chemische Beschaffenheit von Cytoplasma und Zellkern. Ber. der D. bot. Gesell. Bd. XI. p. 293. 1893.
ZIMMERMANN.—I. Die Morphologie und Physiologie der Pflanzenzelle. Breslau, 1887.
—— II. Beiträge zur Morphologie und Physiologie der Pflanzenzelle. Heft I. Tübingen. 1890.
—— III. Idem. Heft II. 1891.
—— VI. Idem. Heft III. 1893.
—— IV. Eine einfache Methode zur Sichtbarmachung des Torus der Hoftüpfel. Zeitschrift f. w. Mikroskopie. Bd. IV. p. 216.
—— V. Botanische Tinktionsmethoden. Zeitschr. f. w. Mikroskopie. Bd. VII. p. 1.
—— VII. Microchemische Reactionen von Kork und Cuticula. Ib. Bd. IX. p. 58. 1892.
—— VIII. Ueber die Fixirung der Plasmolyse. Ib. Bd. IX. p. 184. 1892.
ZOPF.—I. Die Pilztiere oder Schleimpilze. Schenk's Handbuch. Bd. III. Hälfte 2. p. 1.
—— II. Die Pilze. Ibid. Bd. IV. p. 217.
—— III. Ueber einen neuen Inhaltskörper in pflanzlichen Zellen. Ber. d. D. bot. Gesellschaft. 1887. p. 275.
—— IV. Ueber das mikrochemische Verhalten von Fettfarbstoffen

und Fettfarbstoff-haltigen Organen. Zeitschr. f. w. Mikrosk. Bd. VI. p. 172.
ZOPF.— V. Ueber Pilzfarbstoffe. Bot. Zeitg. 1889. p. 53.
—— VI. Zur physiologischen Deutung der Fumariaceen-Behälter. Ber. d. D. bot. Gesellsch. 1891. p. 107.
ZWAARDEMACKER, H.—I. Flemming's Safraninfärbung unter Hinzuziehung einer Beize. Zeitschrift f. w. Mikroskopie. Bd. IV. p. 212.

INDEX.

Abnormal plasmolysis, 239, 241.
Abrus precatorius, 109.
Absorption spectrum, 101, 103, 104, 105, 106.
Acanthospheres, 224.
Acetic acid, as reagent, 60, 63, 71, 90, 94, 99, 113, 188, 220, 252.
—— for maceration, 6.
—— Specific gravity, 262.
Achromatic figures, 194.
Achyranthes Verschaffelti, 205.
Acids, 70, 85.
Acid alcohol, 181.
—— for maceration, 6.
Acid fuchsin, 148, 191, 192, 195, 196, 197, 202, 205, 207, 213, 218, 246.
Aconitine, 120.
Active albumen, 243, 244.
Æthalium septicum, 134, 237.
Agar-agar for attachment, 39.
Agaricus armillatus, Pigment of, 113.
Agave, 210.
Agave americana, 72, 150, 152, 208.
Aggregation, 241.
Albumen for attachment, 40.
Alcanna tinctoria, 71.
Alcannin, as reagent, 71, 74, 89, 90, 91, 95, 209, 210, 211.
Alcohol, as reagent, 51, 69, 82, 124, 125.
—— for dehydrating, 12, 32.
—— for fixing, 176, 252, 253.
—— Specific gravity, 261.
—— Table for diluting, 262.
Alcohols, 69.

Aldehydes, 86.
—— as reagents, 131.
Aleurone, 215.
Algæ, Study of, 1, 5.
Alkaloids, 119.
Alkyl thiocarbimides, 81.
Allium, 81, 88.
Alloxan, as reagent, 131.
Allyl sulphide, 81.
Allyl sulphocyanate, 81.
Aloë, Pigment of flowers, 103.
Aloë verrucosa, 75.
Altmann's acid-fuchsin staining, 195.
Alum, 181.
Alum carmine, 233.
Alum cochineal, 184.
Amido-caproic acid, 82.
Amido-compounds, 82.
Ammonia, as reagent, 87, 94, 96, 113, 122, 123.
Ammonia-fuchsin, 153.
Ammonia, Specific gravity, 261.
Ammonio-magnesium phosphate, 53.
Ammonium, 57.
—— carbonate, as reagent, 67, 88, 117.
—— carminate, 182.
—— chloride, as reagent, 52, 65, 118, 221.
—— molybdate, as reagent, 52, 65, 117.
—— oxalate, 61.
—— —— as reagent, 66, 167.
—— sulphide, as reagent, 121.
—— vanadate, as reagent, 97.

Amorphous starch, 229.
Ampelopsis, 63.
Amphipyrenin, 135, 136.
Amygdalin, 136.
Amylodextrine, 229.
Amyloid, 156.
Angiopteris evecta, 64.
Anhydrite, 66.
Aniline for dehydrating, 17, 29.
Aniline blue, 142, 153, 165, 210, 246, 247.
—— chloride, as reagent, 145.
—— sulphate, as reagent, 86, 145, 147, 157.
—— water, 185.
—— —— solutions, 254.
Anisic aldehyde, as reagent, 131.
Anthoceros, 206.
Anthochlorin, 107, 108.
Anthocyanin, 107, 109, 236.
Antimonic oxide, as reagent, 70.
Araban, 163.
Arabanoxylan, 163.
Arthonia gregaria, 112.
Arthonia-violet, 112.
Artificial cells, Precipitates in, 243.
Artificial precipitates, 242.
Asaron, 85.
Asarum europæum, 85
Ash-skeletons, 168.
Asparagin, 51, 82.
Astrosphere, 198
Atropine, 120.
Attaching sections to slide, 37.
Attractive spheres, 198.
Avena orientalis, 211.
Azolla, 243.

Bacteria, Fixing Methods for, 251.
—— —— by alcohol, 252.
—— —— by heat, 251.
—— —— by lactic acid, 252.
—— Membranes of, 161.
—— Observation of living, 250.
—— Staining, 253.
—— —— spores, 257.
—— —— cilia, 257.

Bacteria, reagent for oxygen, 44.
Bacterio-purpurin, 106.
Bacterium termo, 44.
Balsam-glass, 16.
Balsam Tolu, for mounting, 224.
Barium chloride, as reagent, 49, 59, 62, 70.
Baryta-water, as reagent, 87, 88, 93, 112.
Beale's carmine, 182.
Beggiatoa, 47.
Benzol, as reagent, 124, 127.
Berberin, 120.
Berberis vulgaris, 120.
Berlin blue, as reagent, 132, 168, 172, 190.
—— as stain, 142.
Bertholletia excelsa, 74, 219
Beta, 225.
Betula alba, 70.
Betuloretic acid, 70.
Bitter principles, 99.
Böhmer's hæmatoxylin, 181.
Boletus edulis, 161.
Borecole, 237.
Borodin's method, 49.
Borraginaceæ, 164.
Bottles for reagents, 15.
Brandt's reaction, 97.
Bromine for fixing, 176.
Brucine, 122.
—— as reagent, 51.

Caffeine (see Coffein), 127.
Calcium, 57, 66.
—— carbonate, 60.
—— chloride, 112.
—— —— as reagent, 70.
—— malate, 64.
—— nitrate, as reagent, 70.
—— oxalate, 57, 61, 222.
—— phosphate, 64.
—— sulphate, 62.
—— tartrate, 63.
Callose, 163, 164.
—— Stain for, 165.
Callus, 163.

INDEX.

Calycin, 99.
Calycium chrysocephalum, 99.
Campanula trachelium, 195.
Canada balsam, for mounting, 11, 16, 40.
—— —— for sealing, 43.
Canarin for staining, 10.
Candollea adnata, 194.
Cane-sugar, 78.
—— —— as reagent, 130.
Canna, 225.
Canna Warszewiczii, 205.
Carbazol, as reagent, 145.
Carbohydrates, 75.
Carbol-fuchsin, 215, 254, 256, 257.
Carbon bisulphide, as reagent, 71, 102.
Carbonization, 174.
Carmalum, 183.
Carmine, 182.
—— Beale's, 182.
—— P. Mayer's, 183.
Carminic acid, 183.
Carotin, 101, 106, 204, 209.
Caustic potash for clearing, 9.
—— —— for maceration, 7.
Caustic potash, for reagent, 59, 63, 73, 77, 78, 84, 87, 88, 92, 94, 95, 103, 110, 112, 113, 114, 125, 130, 137, 151, 209, 217, 233.
—— —— for swelling, 8.
—— —— Specific gravity, 261.
—— soda, as reagent, 139, 141, 164.
Caulerpa, 168.
Celloidin blocks, Attaching, 36
—— —— Cutting, 36.
—— —— Hardening, 37.
—— for attachment, 28, 38.
—— for imbedding, 36.
Cell-sap, Bodies in, 240.
—— —— Reactions of, 236.
Cellulin grains, 231.
Cellulose, 138, 139, 149.
—— bodies, 231.
—— Stains for, 142.
Cell-wall, 138.
—— —— Development, 168.

Cell-wall, Minute Structure, 170.
Centrosome, 198.
Centrospheres, 198.
—— Stain for, 199.
Cerasus lusitanica, 137
Ceric acid, 151.
—— sulphate, 126.
Chætopteris, 214.
Chara, 225.
Characeæ, 224.
Chemical differences in walls, 173.
Chlamydomonas, 215.
Chloral carmine, 184.
Chloral hydrate, for clearing, 9, 10, 60.
—— —— for reagent, 71, 90, 193, 227, 233.
—— —— gelatine, 42.
Chlorine for fixing, 176.
Chloroform, as reagent, 71, 102, 150, 151.
Chloroiodide of zinc, as reagent, 110, 139, 140, 143, 155, 166, 245, 246.
Chlorophyll, as reagent, 151.
Chlorophyll-grains, 136.
—— -green, 101.
—— -yellow, 101.
Chloroplastin, 135, 136.
Chloroplasts, 201.
Chlororufin, 106.
Chromatic figures, 193.
Chromatin, 134, 136.
—— spheres, 191.
Chromatophores, 201.
—— Inclusions of, 204.
—— Minute Structure, 203.
—— Methods of study, 5, 201.
—— Pigments of, 100.
Chrome-yellow, 159.
Chrom-formic acid, 178.
Chromic acid for fixing, 177, 240.
—— —— for maceration, 7.
—— —— for reagent, 54, 113, 116, 125, 257.
—— —— for swelling, 8.
Chromic-acid-platinum-chloride, 180.

Chromoplasts, 201.
Chrom-osmic-acetic acid, 178.
Chrom-osmic acid, 235.
Chrysophanic acid, 88.
Chylocladia, 212.
Cicer arietinum, 128.
Cilia, Stains for, 214.
—— of Bacteria, Stains for, 257.
Cinchonaceæ, 210.
Cinchonamin, as reagent, 51.
Cinnamic aldehyde, 146.
—— —— as reagent, 131.
Citrus Aurantium, 94.
—— *medica*, 58.
—— *vulgaris*, 58.
Cladophora, 176.
Cladothrix, 68.
Clearing media, Chemical 9
—— —— Physical, 11.
Clearing sections in celloidin, 38.
Clivia nobilis, 152.
Closterium, 62, 68.
Clove-oil for clearing, 14, 15, 33.
Cloves, 84.
Cochineal, Czokor's, 184.
Cocoa-bean, 72, 126.
Coffee-bean, 73, 94.
Coffee-tannin, 94.
Coffein, as reagent, 242, 243.
Colchicine, 122.
Colchicum officinale, 122.
Coleus, 249.
Collodion, for attachment, 37, 39.
Coloring matters, 100.
Combretaceæ, 210.
Comparison of thermometers, 260.
—— of weights and measures, 259.
Concentration of alcohol, 13, 28.
Confervaceæ, 68.
Congo red, 143, 169.
Coniferæ, 92, 142.
Coniferin, 92, 144, 146.
Conjugatæ, Sheaths of, 157.
Constitution of resting nucleus, 191.
Convolvulus tricolor, 205.
Copper sulphate, as reagent, 130, 137.

Corallin, as reagent, 155, 164.
Cordiaceæ, 210.
Cork, 149, 152.
Corrosive sublimate for fixing, 179, 213, 218.
Corydalin, 122.
Cosmarium, 206.
Cover-glass preparations, 251.
Creosote for clearing, 29.
Crocin, 96.
Crocus vernus, 242.
Cruciferæ, 95, 137.
Crystalloids, Staining, 195.
Crystals, Observation of, 43.
Crystals of ammonio-magnesium phosphate, 53.
—— —— asparagin, 51, 69, 83
—— —— berberin, 121.
—— —— calcium oxalate, 58, 60.
—— —— —— sulphate, 66.
—— —— —— tartrate, 63, 7
—— —— dulcite, 69.
—— —— gypsum, 62.
—— —— hesperidin, 94.
—— —— piperine, 125.
—— —— proteïn-grains, 221.
—— —— saltpeter, 51, 69, 83.
—— —— silver chloride, 48.
—— —— sulphur, 47.
——, Preparation of, 43.
Crystal Systems, 263.
Cucurbitaceæ, 246, 249.
Cucurbita Pepo, 168, 248.
Culture slide for Algæ, 3.
Cuprammonia, as reagent, 139, 140, 155, 157, 162.
Cuprammonia for swelling, 8.
Cupric acetate, as reagent, 90, 115.
—— sulphate, as reagent, 77, 78.
Curcuma amata, 107.
Curcumin, 107.
Cuticle, 148, 150, 152.
Cyanin, 46, 72, 90, 152, 154, **209**, 211, 214, 238.
Cyanophilous nuclei, 191.
Cyanophyceæ, 109, 232.
—— Pigments of, 104.

Cyanophycin-grains, 233.
Cycas circinalis, 58.
Cynoglossum, 164.
Cystoliths, 61.
Cytisine, 123.
Cytisus Laburnum, 123.
Cytoplasm, Bodies in, 240.
Cytoplastin, 135, 136.
Czaplewski's stain for tubercle-Bacilli, 256.

Dahlia, stain, 189.
Dahlia variabilis, 79, 82, 83, 86.
Dammar lac for mounting, 18
Datisca cannabina, 93.
Datiscin, 93.
Daucus Carota, 101, 203.
Dehydrating vessel, Schulze's, 12.
Dehydration, 11.
—— by alcohol, 12.
—— by drying, 17.
—— Klercker's method, 12.
—— Overton's method, 13.
Delafield's hæmatoxylin, 180.
Delicate objects, To mount, 15, 18.
Dermatosomes, 174.
Desmanthus plenus, 234.
Desmidiaceæ, 17, 157, 160.
Development of cell-wall, 168.
Dextrine, 80.
Dextrose, 77.
Diastase, 229.
Diatomaceæ, Pigments of, 105.
Diatomin, 105.
Dicranochæte reniformis, 207.
Digestive fluids for chromatin, 192.
Dionæa, 223.
Diphenylamine as reagent, 50, 69, 83.
Discocrystals, 229.
Dishes for staining, 24.
Double staining, 147.
Dracæna, 210.
Draining boxes for washing, 22.
Drosera dichotoma, 223.
—— *rotundifolia*, 241.
Dulcite, 69.

Eau de Javelle, for bleaching, 6.
—— —— —— for clearing, 10.
—— —— —— for reagent, 110, 143, 152, 228 233.
Ehrlich's solutions, 254.
Elaioplasts, 209.
Ellagic acid, 86.
Elymus giganteus, 211.
Emodin, 87.
Emulsin, 136.
Eosin, 153, 154, 165, 185, 199, 216, 218, 233, 246.
Eosin-hæmatoxylin, 199.
Epiphyllum, 222.
Equisetum arvense, 204.
—— *hiemale*, 54.
Erysipheæ, 231.
Erythrophilous nuclei, 191.
Eternod's apparatus, 25.
Ethereal oils, 89.
Eugenol, 84.
Euglena acus, 230.
—— *Ehrenbergii*, 230.
—— *Spirogyra*, 230.
Euphorbia, 225.
—— *caput-medusæ*, 64.
Evonymus japonicus, 69.
Exclusion of Bacteria from culture, 4.
Eye-spot, 209.

Fats and fatty oils, 71.
Fehling's solution, 77, 78, 92.
Ferments, 136.
Ferric acetate, as reagent, 115.
—— chloride, as reagent, 91, 93, 94, 95, 98, 115, 132, 143.
Ferrous sulphate, as reagent, 45, 95, 98, 115, 219.
Fibrosin-bodies, 231.
Ficus elastica, 61, 62.
Fixing-fluids, Removal of, 22.
Fixing, Methods for, 20, 21, 27.
Fixing-methods for cell-contents, 176.
Flemming's fixing-fluid, 178.
Florideæ, 230

Floridea, Pigments of, 103.
Floridean starch, 230.
Fluids for study of living cells, 4.
Fœniculum officinale, 162.
Frangulin, 93.
Fuchsin, 147, 188, 191, 203, 204, 258.
Fucus, 176.
Fundamental mass of protein-grains, 216.
Fungus-cellulose, 160, 232.
Fungus-gamboge, 91.
Fungi, Study of, 1, 5.
Funkia, 210.

Gaertneraceæ, 210.
Galactose, 163.
Garlic oil, 81.
Gelatinized walls, 154.
Gelatinous sheaths of *Conjugatæ*, 157.
Gentian violet, 148, 153, 185, 186, 255.
Globoids, 219.
Glœocapsa, 110.
Glœocapsin, 110.
Glucose, 77.
Glucosides, 92.
Glycerine for clearing, 11.
—— for dehydrating, 13.
—— for mounting, 41.
Glycerine and chrome alum for mounting, 41.
Glycerine-gelatine for mounting, 41, 42.
Glycogen, 80.
Gold-chloride, as reagent, 81, 122, 126, 127.
Gold-size for sealing, 43.
Gonium pectorale 214.
Gramineæ, 54, 210.
Gram-Günther method, 255.
Gram's method for staining, 185, 255.
Grana of chromatophores, 204.
Granula, 213.
Gratiola officinalis, 222.
Grenacher's borax-carmine, 182.
—— hæmatoxylin, 180.
Growth of cell-wall, 168.

Guiacum officinale, 58.
Gums, 154.
Gypsum, 59, 62, 63, 64, 65, 66.

Hæmatein, 181.
Hæmatochrome, 106.
Hæmatococcus, 106.
Hæmatoxylin, 142, 153, 180, 197, 208.
Hanging drop culture, 2.
Hebeclinium macrophyllum, 63.
Hedera, 205.
Helianthus annuus, 74.
Helichrysin, 109.
Helichrysum, 109.
Hemicelluloses, 161.
Hepaticæ, 210.
Hesperidin, 93.
Higher plants, Study of, 5.
Hoffmann's reagent, 130.
Hofmann's blue, 224, 247.
—— violet, 245, 247.
Humulus, 249.
Hydrocarbons, 88.
Hydrocellulose, 142.
Hydrochloric acid, 48.
—— —— as reagent, 58, 65, 67, 84, 86, 90, 110, 113, 121, 126, 127, 137, 194.
—— —— Specific gravity, 261.
Hydrofluoric acid, as reagent, 53, 55.
Hydrogen peroxide, 45, 117, 178.
Hydrolysis, 139, 162, 163.
Hydroxylamine, as reagent, 147.

Imbedded objects, Attachment to carrier, 35.
Imbedding in celloidin, 35.
—— —— paraffine, 31.
Impatiens Balsamina, 156.
—— *parviflora*, 148.
Inclusions of chromatophores, 204.
—— of nucleus, 194.
India-ink, Use of, 158.
Indol, as reagent, 145.
Inorganic Compounds, 44.
Intercellular substance, 166.

Inulin, 78.
Invert-sugar, 78.
Iodine, as reagent, 155, 227.
—— for fixing, 27, 176.
—— for removing sublimate, 179.
Iodine in sea-water for fixing, 212.
Iodine and potassium iodide for fixing, 214, 224.
—— —— —— as reagent, 80, 102, 103, 106, 120, 122, 123, 128, 129, 185, 186, 233.
—— and sulphuric acid, as reagent, 110, 139, 140, 143.
Iodine-calcium chloride, as reagent, 141.
Iodine-green, 188, 202.
Iodine-phosphoric acid, as reagent, 141.
Iridescent plates of Algæ, 212.
Iridous chloride, as reagent, 124.
Iron, 68.
Isotonic coefficient, 238, 263.

Javelle water, see Eau de Javelle.
Juglans regia, 87.
Juglon, 87.

Karyokinetic figures, 185, 187, 192.
Kinoplasm, 194.

Lactic acid for dried plants, 5.
—— —— for fixing, 252.
Lamellation of wall, 170.
Lathrea squamaria, 225.
Lead acetate as reagent, 93, 94, 98, 107, 109.
Leguminosæ, 162.
Lenzites sepiaria, 92.
Lepidium, 169.
Lepra-Bacillus, 256.
Leptomitus lacteus, 231.
Leptophrys vorax, 230.
Leptothrix ochracea, 69.
Leucin, 82.
Leucoplasts, 201.
Leucosomes, 205.
Lichen-pigments, 110, 111.

Life reagent, 244.
Lignic acids, 144.
Lignified walls, 143.
—— —— Reactions of, 145.
Lignin, 143.
Lilium Martagon, 198, 199.
Lime water, as reagent, 87, 88, 93, 96, 113.
Linin, 135, 136.
Lipochromes, 106.
Lipocyanin, 106.
Lithospermum, 164.
Live-staining, 119, 189, 224, 235.
Living Bacteria, Study of, 250.
Living tissues, Staining, 19, 119.
Loeffler's blue stain, 253.
Lœw-Bokorny reagent, 244.
Lophospermum scandens, 195.
Lupinus luteus, 162.

Maceration, 6.
Madder dye, 95.
Magnesium, 67.
—— oxalate, 67.
—— phosphate, 67.
—— sulphate, 52, 65.
Mandelin's reaction, 97.
Mannose, 162.
Marattiaceæ, 63.
Masked iron, 68.
Maskenlack for sealing, 43.
Mass-staining, 182.
Measures of capacity, 259.
—— —— length, 259.
Melampyrite, 69.
Melampyrum arvense, 194.
Membranes of Bacteria, 161.
Mercuric chloride, as reagent, 122, 128.
—— iodide, 57.
—— —— for swelling, 8.
—— nitrate, as reagent, 129.
Mesocarpus, 234.
Metadiamidobenzol, as reagent, 86, 157.
Metaxin, 135, 136.
Methylal, as reagent, 124.

Methyl alchol, 171.
—— blue, as stain, 142, 153, 188, 190.
—— —— as reagent, 119.
Methylene blue, 166, 168, 173, 191, 192, 224, 235, 248, 256.
—— —— Loeffler's, 253.
Methyl green, 188, 190.
—— orange, as reagent, 236, 238.
Methyl violet, 189, 247, 254.
Micrasterias rotata, 62.
Microcosmic salt, as reagent, 67.
Microsomes, 212.
Microtome knife, 31.
Microtomes, 29, 30.
Microtome technique, 29.
Middle lamella, 6, 167.
Millon's reagent, 85, 129, 137.
Mimosa pudica, Glucoside from, 98.
Mimulus Tillingi, 195.
Minute structure of cell-wall, 170.
Moist chamber, 2.
Mordant for cilia of Bacteria, 258.
—— —— spores of Bacteria, 257.
Morphine, 123.
Mounting delicate objects, 15.
—— in air, 43.
—— in balsam, 16.
—— —— —— with alcohol, 12.
—— —— —— —— aniline, 17.
—— —— —— —— drying, 17.
—— —— —— —— phenol, 17.
Mucus-globules, 232.
Musa paradisiaca, 58.
Mustard oils, 81.
Myrosin, 81, 137.

α-Naphtol, as reagent, 76, 79, 145.
Narceine, 124.
Narcotine, 124.
Neottia nidus-avis, 204.
Nephroma lusitanica, 87.
Nerium, 171, 172, 173.
Nessler's reagent, 57.
Nickel sulphate, as reagent, 50.
Nicotine, 128.

Nigrosin, 189.
Nitella, 190, 224, 225.
Nitric acid, 50.
—— —— for fixing, 213.
—— —— for maceration, 6.
—— —— for reagent, 52, 54, 65, 85, 86, 96, 103, 112, 113, 121, 122, 129, 156.
—— —— Specific gravity, 261.
Nostoc, 233.
Nucin, 87.
Nuclear divisions, Fluid for, 4.
Nuclear membrane, 192.
Nucleic acids, 133.
Nuclein, 234.
—— Artificial, 134.
Nucleins, 133.
Nucleolus, 191.
Nucleus, Constituents of, 175.
—— Inclusions of, 194.

Oil-bodies, 210.
Oil-drops, 208.
Oil-formers, 209.
Oil of bergamot for clearing, 39.
Oils, Ethereal, 89.
—— —— To distinguish, 90.
—— Fatty, 71.
—— for clearing, 14.
Ononis spinosa, 90.
Opium alkaloids, 123.
Oplismenus imbecillus, 211.
Orange, stain, 186.
Orcin, as reagent, 79, 84, 86, 136, 137, 145, 157.
Organic compounds, 69.
Ornithogalum, 210.
Orseillin, 199.
Oscillatoria, 104.
Osmic acid, as reagent, 27, 72, 90, 117, 152, 178, 211, 212, 213, 214, 215, 219, 239.
Overstaining, 26.
Oxalic acid, 62, 70.
Oxalic acid for maceration, 6.
Oxygen, 44.
Oxynaphthoquinone, 87.

INDEX. 293

Pæonia, 156, 217, 219.
Palladous chloride, as reagent, 124.
—— nitrate, as reagent, 81.
Pancreatin, as reagent, 133.
Pandorina, 215.
Paniceæ, 67.
Papaver somniferum, 123.
Paracarmine, 183.
Paraffine blocks, To attach, 35.
—— —— To preserve, 35.
—— for imbedding, 32.
—— -oven, 34.
Paragalactan, 162.
Paragalactan-like substances, 161.
Paralinin, 135, 136.
Paramylum, 230.
Paris quadrifolia, 162.
Paspalum elegans, 82.
Passiflora cærulea, 222.
Paxillus atrotomentosus, Pigment of, 113.
Pectic acid, 167.
—— substances, 166.
Pellicle of protein grains, 217.
Penicillium, 48.
Pepsin, as reagent, 133, 134, 252.
Peridineæ, Pigment of, 105.
Peridinin, 105.
Permanent preparations, 40.
Peronosporeæ, 164.
Peziza, 187.
Phacus parvula, 230.
Phæophyceæ, 230.
—— Pigments of, 104.
Phæophycean starch, 230.
Phellonic acid, 149.
Phenol, for clearing, 10, 60.
—— for dehydrating, 17.
—— for reagent, 71, 145, 146.
Phenols, 84.
Phenosafranin, 166, 168.
Phloionic acid, 149.
Phloridzin, 95.
Phloroglucin, 84, 119.
—— as reagent, 80, 86, 144, 145, 147, 157.
Phœnix dactylifera, 162.

Phosphoric acid, 52.
Phospho-molybdic acid, as reagent, 119, 120, 123.
—— -tungstic acid, as reagent, 128.
Photophore, 25.
Photoxylin for imbedding, 37.
Phycocyanin, 105.
Phycoerythrin, 104.
Phycophæin, 104.
Phycopyrrin, 105.
Phyllosiphonaceæ, 231.
Physcia parietina, 88.
Physodes, 214.
Phytelephas, 162, 174.
Phytophysa Treubii, 231.
Picric acid, for differentiating, 182, 188.
—— —— for fixing, 177, 189, 207, 210, 216, 235.
—— —— for reagent, 123.
Picrocarmine, 183, 255.
Picro-nigrosin, 189, 216.
Picro-sulphuric acid, 177, 235.
Pigments, 100.
—— dissolved in cell-sap, 107.
—— —— in oils, 107.
—— Fatty, 106.
—— in the cell-wall, 108, 109.
—— of *Aloë* flowers, 103.
—— —— Chromatophores, 100.
—— —— *Cyanophyceæ*, 104.
—— —— *Diatomaceæ*, 105.
—— —— *Florideæ*, 103.
—— —— lichens, 110, 111.
—— —— *Peridineæ*, 105.
—— —— *Phæophyceæ*, 104.
—— on the cell-wall, 112.
Piperaceæ, 124.
Piperine, 124.
Pirus Malus, 95.
Pistia Stratiotes, 235.
Plant-mucilages, 154.
Plasma-membranes, 238.
Plasmolysis, 238, 243.
Plastin, 134.
Plastoids, 223.
Platinum chloride, for fixing, 180.

Platinum chloride, for reagent, 56, 81, 128.
—— —— -osmic-acetic acid, for fixing, 180.
Pleochroism, 204.
Pleurotænium Trabecula, 159.
Plugge's reagent, 130.
Podocarpus elongatus, 173.
Podosphæra Oxyacanthæ, 231.
Polarized light, 60, 65, 73, 88, 220, 226.
Polygonaceæ, 88.
Polypodium irreoides, 222, 240.
Polyporus hispidus, 91.
Potamogeton, 211.
Potassic-mercuric chloride, as reagent, 122.
—— —— iodide, as reagent, 122, 123, 128.
Potassium, 56.
—— acetate, as reagent, 70.
—— bichromate for antiseptic, 4.
—— —— for differentiating, 182, 192, 197.
—— —— for reagent, 116, 123, 219.
—— -bismuth iodide, as reagent, 123.
—— -calcium iodide, as reagent, 123.
—— carbonate, as reagent, 86.
—— caryophyllate, 84.
—— chlorate for maceration, 6.
—— chromate, 4, 122.
—— ferrocyanide, as reagent 68, 132, 135, 172, 190, 192.
—— hydrate. See Caustic potash.
—— iodide, 45, 57.
—— myronate, 81, 95.
—— nitrate, 51.
—— —— as reagent. See Saltpeter.
—— oxalate, Acid, as reagent, 70.
—— permanganate for differentiating, 200.
—— platinum chloride, 56.
—— sulphate, 49, 50.
—— sulphocyanide, as reagent, 68, 123.
Primulaceæ, 156.
Proteïds, 128.

Proteïds, Reactions of, 129.
Proteïn crystalloids, 194, 205, 217, 222.
—— grains, 215.
Proteosomes, 244.
Protoplasm and cell-sap, 174.
Protoplasm, Reactions of, 237.
Protoplasmic Connections, 245.
Prunus Lauro-cerasus, 136.
Pulmonaria officinalis, 236.
Pulverization methods, 174.
Pyrenin, 135, 136.
Pyrenoids, 206.
Pyroligneous acid, 180.

Quercus Suber, 149.
Quinones, 87.

Raphides, 58, 60.
Raspail's reagent, 138, 224.
Reactions of cell-sap, 236.
—— of protoplasm, 237
Reagent-bottles, 15.
Remijia purdieana, 51.
Removing air from tissues, 5.
Replacement of alcohol, 14.
Resedaceæ, 137.
Reserve-cellulose, 162.
Resin, 211.
Resins, 90.
Resorcin, as reagent, 86, 145.
Resting nucleus, Recognition of, 190.
Retinic acids, 91, 92
Rhabdoids, 223.
Rhamnus frangula, 87, 93.
Rhodospermin, 223.
Ricinus, 117, 118, 130, 216, 218, 219.
Rochelle salt, as reagent, 77.
Ruberythric acid, 95.
Rubiaceæ, 210.
Rubia tinctorum, 95.
Rutin, 96.

Saccharose, 78.
Saffron-yellow, 96.
Safranin, 148, 152, 185, 186, 207.
Salicin, 96.
Salicylic aldehyde, 146.

INDEX.

Salicylic aldehyde for fixing, 202.
—— —— for reagent, 131, 132.
Salt, 238.
Saltpeter, 5, 240, 263.
Sapindaceæ, 210.
Saponaria officinalis, 230.
Saponification of fats, 73.
Saponin, 96.
Sapotaceæ, 210.
Saprolegniaceæ, 231.
Schulze's dehydrating vessel, 12.
—— macerating mixture, 6, 151.
—— settling cylinder, 16.
Schweizer's reagent, 140.
Sculpturing of wall, 170.
Scytonema, 110, 233.
Scytonemin, 110.
Sealing media, 43.
Secale cereale, 148.
Section-finder, 25.
Selenic acid, as reagent, 122.
Seminin, 162.
Seminose, 162.
Setaria viridis, 67.
Settling cylinder, Schulze's, 16.
Sieve-plates, 246, 248.
Sieve-tubes, Callus of, 163, 165.
—— —— Contents of, 248.
Silica skeletons, 54, 168.
Silicic acid, 53.
Silvering, 173, 226.
Silver nitrate, as reagent, 48, 81, 85, 173.
Silver-solution, 244.
Sinapine, 125.
Skatol, as reagent, 145.
Small objects, Fixing and Staining, 27.
Smilax, 225.
Soda solution, 165.
Sodium, 56.
—— carbonate, as reagent, 113.
—— carminate, 183.
—— hydrate, see Caustic soda.
—— metatungstate, as reagent, 122.
—— phosphate, as reagent, 67, 217, 218, 220.
—— selenate, as reagent, 97, 124.

Sodium silico-fluoride, 55.
—— tungstate, as reagent, 117.
—— -uranyl acetate, 56.
Solanin, 97.
Solanum tuberosum, 202, 222.
Soluble blue extra 6B, 165.
Soluble starch, 229.
Solutions, Percentage Composition, 260.
—— Preparation of, 261.
Solvents for fats, 71.
Sorbus aucuparia, 204.
Specific gravity of solutions, 260.
Spergula vulgaris, 99.
Spergulin, 99.
Sphærocrystals, 63, 64, 65, 73, 74, 79, 82, 85, 93.
Spirogyra, 18, 45, 115. 118, 131, 132, 206, 238, 240.
Spores of Bacteria, Staining, 257.
Staining attached sections, 38.
—— in mass, 24, 182.
—— *intra vitam*, 119, 189, 224, 235.
—— living tissues, 19.
—— Methods for, 20, 24, 27.
—— methods for cell-contents, 180.
—— sections, 25.
Stains for cellulose, .42, 143.
—— —— cuticle and cork, 152, 154.
—— —— lignified walls, 147, 148.
Stapelia picta, 52, 53, 64.
Starch-grains, 225.
—— —— Medium for, 19.
Starch, Recognition of, 227.
Starch skeletons, 229.
Staurastrum bicorne, 159.
Stender dishes, 24.
Stratification of cell-wall, 170.
—— of starch-grains, 226.
Striation of cell-wall, 170.
Strontium nitrate, as reagent, 49.
Strychnine, 125.
Strychnos nux-vomica, 122, 126.
Suberic acid, 149.
Suberin, 148, 150.
Suberized membranes, 75, 148, 150, 152.

Sulphur, 47.
—— compounds, 81.
Sulphuric acid, 49.
—— —— for maceration, 7.
—— —— for reagent, 54, 59, 62, 64, 65, 66, 67, 68, 70, 79, 81, 84, 86, 88, 91, 92, 96, 97, 98, 99, 102, 103, 105, 106, 107, 112, 113, 120, 122, 124, 127, 130, 140, 164, 246.
—— —— for swelling, 8.
—— —— Specific gravity, 261.
Sulphurous acid for washing, 22, 177.
Swelling, 8.
Syringa vulgaris, 98.
Syringin, 98.

Tables for reference, 259.
Tannic acid, 242.
Tannin, as reagent, 45, 143.
Tannins, 114, 242.
Tannin-vesicles, 234.
Tartaric acid, as reagent, 70, 119.
Terpenes, 90.
Thallin sulphate, as reagent, 49, 145, 146.
Theine, 127.
Thelephora, 91, 112.
Thelephoric acid, 112.
Theobromine, 126.
Thermometer scales, 260.
Tholuidendiamine, as reagent, 145.
Thymol, as reagent, 76, 79, 86, 145, 157.
Titanic acid, as reagent, 124.
Tradescantia, 200.
—— *albiflora*, 214.
—— *discolor*, 58, 205, 206, 239, 243.
—— *virginica*, 46, 189.
Trametes cinnabarina, 91, 113.
Trianea bogotensis, 46.
Tropæolaceæ, 137.
Tropæolum majus, 156.
Trypsin, 192.
Tubercle-bacilli, Staining, 256.
Tulipa suaveolens, 242.
Turnbull's blue 169.
Tyrosin, 85.

Unequal water-content of cell-walls, 171.
Unverdorben - Franchimont reaction, 90.
Uranyl-acetate, as reagent, 56, 67, 70.
Uranyl-magnesium acetate, as reagent, 56.
Urceolaria ocellata, 112.
Urceolaria-red, 112.
Urticales, 164.

Vanda furva, 222.
Vanilla, 86.
—— *planifolia*, 209.
Vanillin, 86, 144, 146.
—— as reagent, 84, 131.
Venetian turpentine for mounting, 18.
Veratrine, 127.
Veratrum album, 127.
Vessel for staining sections, 26.
Vesuvin, 165.
Vicia Faba, 46, 187.
Vinca, 172, 173.
Violaceæ, 137.
Vitis Labrusca, 63.
—— *vinifera*, 220, 222.

Washing, 22, 27.
—— apparatus, 23, 27.
Wax, 74.
Wax feet for cover-glass, 1.
Weights, Table of, 260.
Willows, 96.
Wood-gum, 144.
Wound-gum, 157.

Xanthein, 107.
Xanthin, 103, 106, 209.
Xanthine, 128.
Xantho-proteic acid, 129.
Xanthotrametin, 113.
Xylol for clearing, 14.
—— for imbedding, 33.

Zinc chlorine, as reagent, 231.
—— sulphate, as mordant, 199.
Zygnema, 115, 176, 206, 235.
Zygnemaceæ, 157, 158, 234.

www.ingramcontent.com/pod-product-compliance
Lightning Source LLC
Chambersburg PA
CBHW022057230426
43672CB00008B/1204